Against All Odds

Against All Odds

My Life of Hardship, Fast Breaks, and Second Chances

Scott Brown

An Imprint of HarperCollins*Publishers*

HarperCollins books may be purchased for educational, business, or sales promotional use. For information please write: Special Markets Department, HarperCollins Publishers, 10 East 53rd Street, New York, NY 10022.

Scott Brown is a lieutenant colonel of the Army National Guard. Use of his military rank, title, and photograph in uniform does not imply endorsement by the Department of the Army or the Department of Defense.

An extension of this copyright page appears on pages 461–62.

The poem on pages 145–46 is from *Hand-Book of Wakefield*, by Will E. Eaton, 1885.

FIRST HARPERLUXE EDITION

HarperLuxe™ is a trademark of HarperCollins Publishers

Library of Congress Cataloging-in-Publication Data is available upon request.

ISBN: 978-0-06-201791-8

11 12 13 14 ID/OPM 10 9 8 7 6 5 4 3 2 1

To Gail, and to Ayla and Arianna,
forever and for everything

CONTENTS

Against All Odds

PROLOGUE

By the time I turned eighteen, I had moved seventeen times and lived in at least twelve different homes. Most were rental apartments, second-floor walk-ups in slightly sad and dilapidated converted houses, where walls had been added and the rooms and floors chopped up one by one.

My bed, when I had one, was invariably under the hard slope of the eaves. I also made do with couches and cots, and there was a time when my mother, my sister, and I all slept in the same small room.

When I was twelve, our apartment's backyard stood at the edge of a thick tree line and was so damp and dark that the dirt ground stayed bare all year round. Of the five houses I knew, one was a doll-sized rental in the backyard of another home and the other four belonged to relatives or to whatever man my mother happened to be married to at the time. We were visitors there; they were never our own.

At school, I was often a free-lunch kid, ravenous for whatever hot food came out of the cafeteria line. Constrained by her choices, good and bad, my mother worked hard, often at multiple jobs, to keep a roof over us, put clothes on our backs, and pay babysitters, and she bought food and a few extras with whatever was left over. I remember days when the largest things we had in our fridge were milk and blocks of yellow government-issue cheese.

My dad was largely gone from my life before I turned one. He materialized only on rare weekends, a smooth talker with his foot on the gas and the convertible top down. A couple of times when I ran away, he was my destination, but even then, I never stayed long.

Looking back, what saved me was basketball—a game, ironically, that was homegrown, invented in the Western Massachusetts city of Springfield in 1891 by a YMCA instructor who was looking for a way to keep his gym class busy during a bout of rain. He started with a peach basket, a soccer ball, and borrowed rules from a kids' game known as "duck on a rock." I doubt

Dr. James Naismith could have pictured me, as a kid of twelve, thirteen, or fourteen, riding my bike several miles after a snowstorm with a ball cradled under one arm and a snow shovel clutched in my hand to clear off the courts so I could shoot hoop. Not just for ten minutes, but for hours on end, until my fingers got so numb that I could no longer feel the ball balanced between them.

What saved me too were my friends, my teammates, my coaches, and even the cops and a judge, and later the military, although I didn't quite know it all then. I could have easily been the kid with the rap sheet and the long record, rather than the accolades and the record high scores.

I look back on my life now, though, and I can honestly say that there isn't one thing I would change: not the arrest, not the violence, not the hunger, not the beatings and the brute struggles, not even cleaning up someone else's vomit in the stairwell of my dorm at Tufts for $10 in quick cash from the resident adviser because I had no money for extra food. I wouldn't change my decision to pose for *Cosmopolitan* magazine, which helped to pay my way through law school, forced me to grow up even more quickly, ultimately led me to meet my wife, and also slowly led my father back toward me.

Whatever the widest boundaries are for a Wrentham selectman, a Massachusetts state senator, or a United States senator, I am sure that my life lies outside them. But I wouldn't change any of it, because while it was too often hell as a boy—and I myself was at times a hellion—those years and that life made me the man I am today. I hope, too, they made me a better man.

Chapter One

BUSTED

The Liberty Tree Mall was our last stop. It sits right off Route 128 in Danvers, Massachusetts, its big anchor stores rising up flat and square, like stackable Lego blocks. At one end was a Sears with tools and tires, appliances and overalls, and at the other, a Lechmere store, with displays of shiny new luggage, sporting goods, and jewelry, as well as an electronics section and, most important, a record department. We pulled into the lot, away from the hum of highway traffic headed south toward Canton or Braintree, or looping around toward Boston itself. One of my good friends was riding shotgun in the car; one of his buddies was driving. Both were a couple of years older than me and both were basketball players. I was sitting in the backseat. I was thirteen, a few months shy of fourteen, but I was already closing in on five foot eleven. My hair hung long, skimming over my shoulders, drooping into my eyes.

I was lucky in that moment, not lucky that I was along for the day—I had been hanging around these guys for a couple of years, shooting hoop, going to their parties, sipping their beer. That afternoon, I was lucky I wasn't the one driving.

We parked, rolled up the windows, hit the locks of the car, and then shuffled across the baking asphalt. The air was hot, that sticky, humid July heat, where the sky turns thick and white and presses back down upon you until each breath seems liquid, like sucking pool water into your lungs. The weather was why we weren't on the basketball court; another reason was that when both guys woke up in the morning, they had decided that they wanted some records. I wanted some too. I had a few records, but all my friends owned dozens and dozens.

We had already been to two record stores that day in another mall, but there was one more inside the sprawling sections of Lechmere, beyond the luggage displays and jewelry counters that beleaguered husbands crowded around when it got close to the holidays. We ambled through the store in the air-conditioned cool, beneath the bright fluorescent lights, which made it impossible to tell afternoon from evening. I had on overalls, blue-and-white railroad stripes with a big front placket. I called them farmer pants, but my mom or I had most likely found these in a surplus store or a discount bin. On top, I wore my junior high basketball jacket, a bright red nylon with a heavy lining for the damp, bleak Massachusetts winters. It had our team's emblem, the Wakefield Warrior, a big Indian chief in profile with a full feather headdress, stamped across the front. If there was a moment when life became premeditated, it was when I got dressed that morning.

We walked into the store and went over to look at the music, arranged alphabetically, *A* for America, *B* for Beatles or Bee Gees, *C* for Creedence Clearwater Revival, *D* for the Doors. There were the small 45s with one song on either side, but we wanted albums. Although the radios played Elton John, Steely Dan, and the Steve Miller Band, our tastes ran to hard, searing guitar rock, like Black Sabbath, Led Zeppelin, and Deep Purple, or the pounding, mournful songs of Jim Morrison. Morrison's "Riders on the Storm" echoed in my mind. I followed, and after they had thumbed through the stacks, I noted what they had chosen, what was cool. My friends headed off to another area of the store, but I stayed behind. After checking around, I leaned in, unfastened the two side buttons on my overalls, and slipped an album behind the jacket. Then another and another, and another after that. I could comfortably carry five. The cellophane covers slid easily against each other, and the thick mass came to rest on my stomach. Trying to look nonchalant, I popped the metal buttons back through their holes, zipped the jacket, and began to amble out of the store. The other guys had already left, and they were waiting for me in the parking lot. I was almost to the doorway; perhaps I was even grinning.

Suddenly a man's hand reached out and patted my back. Instinctively, I stopped and that same man said, "Hey, it's awfully hot out today." I tilted my head, which was angled down, and looked out through the thick fringe. He was wearing regular street clothes, but still my heart began to pound. "Yeah, it is," I managed to reply.

"What are you doing?" he asked me. I mumbled, "Just hanging out." Then the man's hand slipped around my shoulder and gave three hard pats on my stomach. There was no mistaking the clean-cut cardboard edges or the hard feel of the album covers. "I'm store security," he said. "Why don't you come with me?" I had the sickening feeling that he had probably been watching me, in my winter jacket, the entire time.

He led me out through a side door, away from the bright lights, to the concrete back corridors of the mall, where everything was gray cinder block and the ceiling lights were skinny tubes that flickered and hummed. I had never been in the bowels of a store before, where stock was rolled on dollies off the loading docks, where employees entered and exited, and dull doors opened into backroom offices. I walked wordlessly, head down, afraid that I might see someone I knew. The soles of my sneakers made squeaking noises on the hard, flat floor. He led me into one small room, which housed the security office. It was sparsely furnished, with nothing more than a metal desk, an industrial chair, and a telephone. My friends were still waiting for me in the parking lot, having no idea that I'd been caught. The guard told me to unzip my jacket, and I removed the records, the bright cover art already peeking up from behind the placket of my overalls. He looked at each album and then began asking me questions, including, "How did you get here?" When I told him that I had gotten a ride, he asked me to take him out to the parking lot, where my friends were leaning against the sides of the car.

Once we reached them, the guard told the driver to open the doors and then the trunk. There were twenty or thirty other records inside, all still tightly wrapped, from other stores where we had stopped earlier that afternoon. I don't remember whose idea it was to boost the records, probably mine, but the other guys had gone along. In that moment, I think I took the blame for everything.

The guard picked up the records and we walked back inside. I returned to the solitary room. He asked me for my parents' phone number, and I gave him my mom's number at work. I didn't even consider giving him my dad's. I never knew on any given day where he was or if he would come.

The security guard called my mother and then he called the cops. The other guys were older, but I was a juvenile, and I had been caught with the records, so it was easier to pin the entire haul on me. At that moment, it wasn't as if I saw my future flashing before my eyes, but I was definitely scared. I was thinking: What about basketball, what about school, what would my punishment be? The guard was lecturing me about stolen property and then I saw the dark blue uniforms of the cops. They looked at me with inscrutable stares, asked some perfunctory questions, examined the albums, and wrote a citation and court summons on a thick pad with layers of carbon paper. I was, they said, remanded to the custody of my mother until my court appearance, two weeks later. My mother came straight from work, her face red, her leather purse clutched like she might at any second smack me with it. I braced myself for the car ride home. Not only had she left work early today; she would have to miss work again to take me to court. And I had been caught stealing. I sat in gloomy silence as she yelled at me the whole way, her hands intermittently flying off the wheel. When I managed to get out a word, she immediately cut me off with another volley of screaming. "How could you do this?" I heard that line again and again.

How could I do this?

I did it the same way that I stole a three-piece suit from Park Snow, the stand-alone department store in downtown Wakefield, because I had nothing to wear to a school dance. I had walked in, carrying a duffel bag, tried on a suit that was right against the wall to make sure that it fit, then inside in the solitary confines of the dressing room, stuffed it in the bag and sauntered out.

And I did it the same way I stole food.

That had started earlier. I was eleven or twelve and hungry all the time. Ravenously hungry, to the point where my stomach would often ache, and I would sit on the couch with my knees drawn up to my chest, as if I could physically shrink the space between my lungs and my abdomen. There were long stretches of hours when my mother was not home; she had office jobs, hospital jobs, and many stints as a waitress. It was only my half sister Leeann and me, and a babysitter who was there mostly for Leeann. In the late afternoons, I would go on my bike, a rusty, secondhand blue Schwinn, down to the A&P in the center of Wakefield, a couple of stores away from Park Snow. Sometimes, I went straight after basketball practice. If it was after practice, I had my gym bag with my sweaty tube socks and clothes. Otherwise, I wore my railroad stripe overalls. I would grab a cart and meander through the store aisles, picking out a loaf of bread, maybe some juice. The mothers with their cranky toddlers or trailing grade-schoolers were too busy to notice when I lingered by the meat case. There, under tight plastic wrap, were piles of freshly ground hamburger and rows of thick-cut steaks. I would pick up one or two packages, examine them, and then pop the button on my overalls and slide them in, or drop them into my duffel, and fumble for a second to slip the sweaty clothes over the top. There was so much meat, I reasoned, how would they miss a package or two? They could never sell it all. No one would buy every last hamburger package. I was saving it from the Dumpster. And I was starving.

Most times, I did not have to suck in my belly to feel each individual rib.

After the meat counter, I would push through the aisles, maybe snag some milk, which I could down by the gallon, and then head past the cereal boxes and white rice to the registers. I always bought some of the cheaper items, but first, I had learned to hang back for a minute and analyze the checkout clerks. I never went to the middle-aged ones; I always chose the line with the teenage kids working after school, kids whom I sometimes knew, who sullenly punched the numbers on the register, who would never look at my now-bulky overalls or gym bag. What sixteen-year-old kid imagines a twelve-year-old stealing food? Certainly not in Wakefield, Massachusetts, a pleasant middle-class town, a commuter town on the railroad line into Boston, a pretty quiet place back in 1971.

I would ride home, clutching the grocery bag in my right hand, feeling the jolt as it bumped against my legs or swung against the wheel. With my left hand, I steered the handlebars. The meat was always in my duffel, slung over my shoulder, messenger style, or tucked safely behind the placket of my overalls. We lived in a second-floor walk-up apartment in a converted house. My room was what had been the attic. It was a long, narrow house, completely unadorned.

The backyard was so dark from woodsy overgrowth that nothing grew and the ground was usually brown. I would ditch my bike in a shed out back and carry in my haul.

I didn't even bother putting the meat in the fridge. I dropped it on the counter, pulled out an old, scratched aluminum pan, and flicked on the range. Sometimes I slit the cellophane with a knife; other times I just tore it with my bare hands. I panfried steaks or burgers, listening to the sizzle of the meat, feeling the quick splatter of hot grease across my knuckles as I flipped them. I learned by trial and error, trying to remember the early cooking lessons from my grandmother, to get the blue flame of the burner just right, so the steaks or burgers would not be charred by the time I turned them, or still be so raw that the outside layer stuck to the bottom. Sirloins were the best, the tenderest cut. Burger meat was hit-or-miss; some bites were gristly with the remnants of tendons and butcher scraps cut too close to the bone.

When at last they were hot and dripping pink juice on the plate, I would cut up the meat or serve the burgers, giving some to Leeann, who was six. She hardly ever asked where it came from or why I was cooking it, but if she did, I would just say it was in the fridge. She would nod and start eating in silence. I ate my own portion in gulps, not bites, until gradually the stabbing sensation in my stomach would give way. Then I cleaned up, scrubbing every pan, every plate, opening the windows even in the winter, wiping the stove, the counter, pumping a quick spray of room freshener, so that there was no trace. Mom would come home to the often barren fridge, her shoulders slumped and aching from a long day at work, and ask what I wanted for dinner, and I would say, with complete truthfulness, that I was OK; I was full.

That seemed to satisfy her. Once in a while she might ask what I ate, and I would mumble about grabbing something after practice. And that was all. There would be the clink of ice in the glass and the splash of vodka, followed by the scrape of a match on the back of a free book from one of the restaurants, bars, or lounges that lined the highway known as Route 1. She always lifted the matches they kept in bowls by the door or dropped in the center of amber glass ashtrays for the patrons. The match head would hiss, and she would light a Marlboro Light and draw in a deep breath and then exhale streams of gray smoke in a long trail out of her mouth and nose. She might ask me about basketball or school. I kept the answers short. Or she might hassle me about dirty clothes or a messy room, and I would match her word for word. Whatever she threw in my face, I lobbed right back. It was the dance we did, with the television droning in the background. She came home tired in a used car to an apartment where there never seemed to be enough cash; she was thirty-four, and in fourteen years had married and divorced three husbands.

Some nights she was itching for a fight, but others she was too tired to notice if I'd eaten or done my homework. Then there were the nights when she was simply gone, out for the evening with her friends to a club or a bar, where they laughed long and loud and left behind lipstick stains on the cigarette butts and liquor glasses. When we were short on rent, she waitressed in one of the places out along Route 1, flirting I think with the middle-aged men who were the ones to get the check and who she hoped would reach in their wallets for a stack of bills, saying, "Here, honey, keep the change." Her other gigs were mostly on the weekends, when she dressed in black and served identical plates of catered food for extra cash at wedding receptions or banquet meetings. If there were uneaten meals, she might take those home in foil tins, extra portions of chicken breasts with congealed mushrooms and cream, rice gone a bit dry, cold tomatoes that hours earlier were baked and bubbling with bread crumbs. The nights when she had a catering job were the nights when I, as I got a little older, was most likely to take the car for a spin. I was twelve when I first slid behind the wheel.

I began by backing the car out of the driveway of our duplex so that I could get to the basketball hoop, just easing it out and angling it along the curb down the street. But gradually I became more daring. Months passed and the distances got longer, until at some point, I was driving. I learned to drive by watching some of my older friends when they picked me up for league basketball games, staring at the way the steering wheel rolled through their hands, how they flicked the turn signals with a quick tap of a finger or a thumb and pressed down effortlessly on the gas or the brake, rocking their large basketball player feet up and down on their heels.

There were even times when I would drive one of the older kids home after he'd knocked back a couple of beers in someone's basement and tossed me the keys. I was tall, I looked old enough to pass for seventeen, and my long hair hid what was left of any little-kid face. Other drivers might glance over, but they rarely glanced twice. I was like any other kid out with a parent's car on a Saturday night. I was always very conscious of following the rules of the road. In retrospect, I was also just plain lucky.

I taught myself to drive in my mom's Chevy Impala,

a bright white car with red vinyl seats and four doors, wide and roomy so everyone could pile in. If I knew she was going to be gone from three until eleven, hitching a ride with her friend, I would take the car out from five to eight, with two or three of my friends in the backseat. In the beginning, to teach myself, I drove the car down to the vacant parking lot outside the American Mutual Insurance building, which backed up against Lake Quannapowitt in Wakefield. It was the same building and the same company where my mother had worked, typing and filing, but the irony was lost on me as I practiced turning, parking, shifting the car into reverse, stopping on a dime. It didn't matter if my mother had her purse with her; I had my own set of keys. One afternoon, I rode my bike down to the local hardware store and had a copy cut at the cost of some pocket change.

At twelve, thirteen, and fourteen, I was not as smooth as the sixteen- or seventeen-year-olds, but I was nearly as tall, with a clear view over the wheel, and I was careful. I didn't drive after I'd tried a beer, and I didn't race or try to beat the light. I only stayed inside the confines of Wakefield. No highways, no busy routes with cars passing and pulling into and out of view. The driving became like a running joke, although I was always careful to put back in just the right amount of gas, so she would never wonder when she came down to start the engine in the morning.

When she drove me home from Liberty Tree Mall, yelling at me for what a disgrace I had been, she never thought for a second that she just as easily could have gotten a call from a Wakefield cop on a Saturday night, someone in a blue uniform who had taken a second look and had busted me for being behind the wheel without a license, or, God forbid, for crashing into, injuring, or even killing someone else while I was driving underage. I never thought of it either.

That night, as I did on most nights, I took my basketball to bed. I would trace my fingers over the black ribs and talk to the ball. It smelled of sweat and dirt and whatever had dropped from the sky or trees onto the court. And it listened companionably in the darkness. I slept with my hand resting alongside its worn pebblelike surface.

I assume Mom called my dad that night or the next morning, if only to say, "Look at what your son has done," but I never heard from him after the incident at the mall. He didn't come around afterward, and there was no father–son talk. His absence was as loud as the proverbial slam of the screen door when I was mad and raced out, wanting to be anywhere but home. After that, I was forbidden to see my older friends, the guys who drove and who had driven me to the mall, but my mother exercised no control over the basketball court, and we still met up there, where we rebounded, took free throws, guarded, and jumped, and never said anything about what had happened on that July afternoon. And we still hung around anyway. I didn't even know if anyone else had gotten a summons, and I didn't want to ask. I had two weeks to wait for my court date at the Essex County Courthouse.

The morning arrived and it was hot. I wore a shirt and tie, and the suit that I had lifted from Park Snow, sweating in the heat. The Impala had air-conditioning, but air-conditioning burned more gas, so we drove to Salem with the windows down.

It was the Salem of the infamous witch trials, where 142 people were accused of witchcraft and 19 were hanged. One man, Giles Corey, a cantankerous farmer, was pressed to death over two days, lying naked on his back under the September sun as one by one stones were placed on his chest until his ribs and lungs were crushed, because he had refused to enter a plea of guilty or not. But I hardly knew any of that. I knew nothing of Salem's days as a port that shipped salt cod to the Caribbean and Europe and as the final destination for bags of sugar and barrels of thick, dark molasses from island plantations, or tins of Chinese tea, arriving on boats that had crossed around the bottom of the world. I was a boy who had at most been to Boston once on a class trip. To tell me that Nathaniel Hawthorne began *The Scarlet Letter* in the Customs House near Pickering Wharf would have done little more than bring a glazed look to my eyes. More appropriate to me that morning was the fact that Salem had been a hotbed of privateers seeking riches on the high seas; Salem ships captured or destroyed some six hundred British vessels in the Revolutionary War and struck again during the War of 1812.

The courthouse was a reminder of that prior world of prosperity and commerce, a beautiful old building sitting in the middle of downtown.

Inside were the sounds of footsteps on its glossy floors and men in suits and ties, clutching briefcases, moving with a purposeful stride. My case, I learned, was going to be heard by Judge Samuel Zoll, who stood six foot four in his flowing black robes. I walked into the courtroom with my mother, with my eyes down. A juvenile representative had been assigned to me. Judge Zoll looked at the defender and my mother and then at me and in a booming voice asked to see me in his chambers, just the senior probation officer and me, alone. I left my mother outside the courtroom and a bailiff escorted us back through the labyrinthine hall. Wordlessly, he ushered me forward, and I walked into the judge's private chambers, my hair too long, my feet shuffling, my palms damp. Behind me, the thick wooden door shut, and across the desk, a single pair of eyes bore down.

Chapter Two

DAN SULLIVAN'S HANDS

My first photo, or the first photo that remains, was taken when I was about six months old. It is a picture of me surrounded by my father's sports trophies, basketball mostly, but maybe a few other sports too. I am sitting in a onesie, a basketball on my lap, amid my father's monuments to glory. In less than six months, the man and his trophies would be gone.

I never got the exact story of how my parents met. In one version, my mother is a cashier and hostess in a restaurant along the thin slip of New Hampshire shoreline, with its clam chowder joints and seasonal souvenir stores, and my father is an Air Force flyboy stationed at the nearby Pease Air Force Base, a World War II landing strip that was later transformed into a spanking new institution designed to wage "cold war." The official Pease Air Force Base is just one year old. He comes in with his buddies, she seats them and rings them up, they talk, they flirt, she gives him her number, and he calls. In another story, she is the runner-up for Miss Hampton Beach, and my father is an Air Force logistics or maintenance guy, an enlisted man, not an officer. His name is Claude Bruce Brown, but he goes by C. Bruce Brown. My mother is Judith, Judy to her friends. She is the younger of two daughters, the captain of her high school cheerleading squad. Her father is an electrical engineer with Boston Edison, a proud graduate of the Massachusetts Institute of Technology (MIT). Her mother keeps the home. Judy is a good student and very artistic, and is planning on attending art school. She has a full scholarship to an art school in Boston. She draws, sketches, and paints and has her creations framed to hang on the wall. But she never went to art school. Instead, with nothing more than a high school diploma, she married my dad.

What I do know for sure is that my dad was handsome, and still is; my mom was beautiful, and still is;

and they met in the summer of 1957. They were like a lit match, sudden, sulfurous, and nothing but flakes of ash and char after they had ceased burning.

They married fast: six months of dating and then straight to the altar, saying their vows in the chapel on Pease Air Force Base. They were not love-struck teenagers; my mom was twenty, and my dad was twenty-one. But they ricocheted down the aisle as if it were a shotgun wedding. They set up house in Portsmouth, close to the air base, but never on the base's actual grounds. Their home was the left side of a clapboard house that my mother's father, my grandfather, owned, up the road from Portsmouth's downtown cluster of sturdy redbrick buildings. Grandpa had split the house down the middle and rented out both ends. He came from New Hampshire, had been born in Portsmouth, and grew up there in a simple, saltbox-style house with no yard to speak of, on a quiet block. When his parents and spinster aunt died, they left him their homes, so he had some modest rental investments in addition to the narrow, single-family home he owned in Wakefield, Massachusetts. If he thought it was a bad sign that his new son-in-law couldn't come up with a place to live on his own, it seems that he held his tongue.

It may have been a love match to start, but after the first few months, my father began staying out later, disappearing to hang out with his buddies, coming home late or making excuses about why he had to spend extra time on the base. In the long nights, odd shifts, and flimsy excuses, there were hints of other women. A year into their marriage, my mother got pregnant; I was born, breech, at the Portsmouth Naval Hospital just over the metal suspension bridge in Kittery, Maine. I assume my father handed out cigars. The date was September 12, 1959.

Pease housed long-range bombers and payload aircraft designed for nuclear strike operations, but I doubt my mother knew or cared. She had a crying newborn and a husband who was AWOL from his own home. The last straw was when she walked into a ladies' dress store downtown with me in tow and heard a group of women giggling and gossiping. One was telling a story about an Air Force man whom she had been seeing. That man, my mother says, was C. Bruce Brown.

Angry and spitting, my mom packed up me and what little she had and left in a hurry. It was about fifty miles from Portsmouth to Wakefield, and my mother no doubt cursed Bruce Brown across every one. My father never acknowledged the other women; he just made references to being too young and being suddenly burdened with too much responsibility, primarily a wife and a baby son. My mother left him behind, but he is the one who vanished like an apparition. He left the Air Force and went into insurance, successfully selling all kinds of policies, big and small. He moved down to Massachusetts, but never close enough to mean anything. I would wait for him for hours on weekend mornings or afternoons, my nose pressed against the glass of the door, my breath making little rings. The harder I pressed, the quicker I might see him coming, catch that first glimpse of his convertible motoring down the block. But more often than not, he didn't come. He was like a mirage in the desert, a picture that I created in my mind, which I could also slice clean through with a wave of my little-boy hand.

There is one other photo of me from that time, about the age when I turned one, a studio shot with a perfect monochrome pearl-gray background in which I am clutching a small white bunny with floppy ears and wearing a checked playsuit with white leather lace-up baby shoes. I am smiling, but it is a tight, nervous smile. I must have been told to smile, to laugh. I do not look as if I want to, but my legs are not sturdy enough yet to do much beyond toddling. It has not occurred to me to run.

When my meticulous grandfather cleaned out the remaining wreckage of the apartment, he found notes from women tucked away in drawers or a closet, loopy handwriting scribbled on pieces of paper or the torn covers of matchbooks with phone numbers and seductive messages. My father stuffed them into his pockets the way a salesman collects business cards out on the road and then empties them out at home.

After my mother packed up and returned to Wakefield, we lived for a time in the home where she grew up, where my grandfather wore ties even under his sweaters and my grandparents never exchanged so much as a chaste public kiss or a hug, although I always knew that they cared deeply for each other. But at night, Bertha and Philip Rugg closed the bedroom doors on twin beds in separate rooms.

I loved my grandparents' home. It was a modest place on the quiet block of Eastern Avenue, with a living room to one side, a dining room to the other, and a kitchen in the back. There was a yard to play in and there were hot dinners at night, pot roast, chicken, and new potatoes boiled and bursting from their split skins. My grandmother could cook anything, from pies to vegetables; the toughest cut of flank steak became tender in her hands. My grandfather was not a talker. Occasionally, he'd point out something he found interesting, but he was a quiet, reticent man, opaque and flinty like his native New Hampshire stone.

He had gone to MIT in Cambridge, Massachusetts, leaving Portsmouth to learn engineering, and even now at home he studied his electrical engineering magazines, worked crossword puzzles, played solitaire, and read the newspaper in his easy chair, the pages rustling as he turned them one by one. Sometimes he would put on the baseball game, and the announcer's voice would rise and fall with the whistle of the pitch and the crack of the bat. Grandpa would watch, and say nothing. Although he took the train into Boston every day for work, I don't think he ever attended a Red Sox game. He was a young man who married at the start of the Depression, and he was thrifty. He didn't really have hobbies, except for solitaire, canasta, and reading books. He spent his remaining free time tending to his rental houses in Portsmouth, where he had the same families living for as long as twenty or thirty years at a time. Over the span of a decade, he barely raised their rents, preferring to keep everything the same. He liked tangible things, like property or a few bank or utility stocks that paid regular dividends.

When my parents were newlyweds, my father had borrowed money from Grandpa to buy a car. He had never paid it back, and although my grandfather never openly spoke ill of my father in front of me, he never forgot that particular debt or forgave him. When I was grown, I borrowed $1,000 from Grandpa and promised to pay $100 in interest (cheap money at a time when most banks were charging 20 percent or so). I repaid the $1,000 and he reminded me about the interest, crossing off the obligation only when the last dollar had been delivered. He was meticulous in his calculations.

Philip Rugg was a civic-minded man, joining the John Paul Jones Society to help refurbish the famous John Paul Jones House in Portsmouth. He was inducted into the local Masons, where he rose to become a parliamentarian. He and my grandmother were Unitarians, and I was dedicated, their version of a baptism, in the Unitarian church and went there many Sundays when I was very young. But when the church grew too liberal for my grandparents, they quietly stopped attending.

Along with his crosswords, he had a passion for putting together puzzles, and there was invariably one spread out on the puzzle table for us to work on. Even as a teenager, I still came over to work the puzzles with him.

My grandmother had been a teacher, and she saw me as her private pupil. She taught me to sew, to knit (though knitting never really took), to iron, to clean, and to cook, including tests on safety, like how to place the pan on the burner, which way the handle should point, or what to do in the event of a grease fire. It was often just the two of us. My grandfather commuted to Boston, my mother was gone at her office or waitress jobs, and so in the daytime, Gram watched me alone in the house on Eastern Avenue.

After some months had passed, my father did start coming around again. I don't really remember his visits—I was too young—but he came over, sometimes to see my mother, sometimes to see me. One Saturday early in the spring, before the trees had leafed out, when it was more mud season than anything else, my mother dressed me up in my little Easter suit and hat and parked me by the door to wait for my father to appear. I had stiff, shiny shoes and she had wet my hair and combed it back under the hat. I waited and waited and my father never showed up. Eventually, my mother must have called or he did, and she said, "Where were you? Scott was waiting for you," with a bit of a nervous screech in her voice. My father replied, "Well, I'm getting married."

"Married?" My mother took it like a sucker punch, a verbal smackdown as good as anything she ever got later from the brute force of another man's fists. After our return to Wakefield, my parents had been talking about reconciling, and from there, my mom had probably made the leap to a full reunion. People did it all the time. Hollywood stars in their glossy romances came together, broke up, and reunited. I didn't know what reconciling was; I only felt the displaced fury of my mom.

In her view, C. Bruce Brown had been stringing her along once again, just as he had two years before in Portsmouth, with his excuses and his pockets stuffed with numbers scrawled on the back of matchbook covers. This time, my mother had been his fallback in case this other woman, Delores, didn't pan out. Or perhaps she was only one on a long list of other women. And the sting soon became greater. Delores was pregnant. My father was going to be a father again, with a new wife and new children. I can only imagine my mother's despair and her desperation at having been abandoned twice. But I lived with the bitterness that resulted. A high school graduate who had rejected art school for the altar, she could lay all her disappointments at the feet of one man, C. Bruce Brown. From that day on, my mother could barely stand to be in the same room with him, and when time or circumstance or their shared son forced them to be face-to-face, it seemed like only moments before the insults flew and the deep needling began. And as his son, I was a daily reminder of him.

Now decidedly stuck in Wakefield, my mother was itching to leave, not the town itself, but the confines of her childhood home. Her high school friends were married, as was her sister, who had her own children and home, while my mother was curling up each night under the covers of her teenage bed, in the same room where she had practiced her cheerleading chants and had dreamed of her senior prom. As soon as she could, she rented an apartment, one of the many anonymous places we would live in—converted houses that had been split into clusters of rooms with efficiency kitchens. One of the first was a rental place on Avon Street, sandwiched between the railroad tracks running along North Avenue and Main Street. I went to nursery school in a red house a few blocks away on the other side of the train tracks. I still remember sitting in the back room and learning on a Friday in November that John F. Kennedy had been shot. I watched tears roll down the teacher's cheeks, and I also cried and cried. At home, on Avon Street, I had a little ring with Kennedy's picture on one side and an American flag on the other, and I twisted it around so Kennedy's face was always pointing in.

Avon Street was where we were living when Dan Sullivan came home.

I don't remember meeting Dan, or going with him to the park or to a diner for a milk shake. It was simply that one day he appeared in our small space, and my mother announced that she was marrying him. They were married quietly in the living room of my grandparents' home on Eastern Avenue by a minister who lived next door. Dan was a truck driver, a short and long hauler of petroleum products for a local oil company. I remember his rough, callused hands with fingernails that were always gray or grimy under the rims. He was a rugged Irish guy and handsome, with hazel eyes, fair hair, and ruddy skin that turned bright red in the heat. In the summers, his arms and neck were almost always burned, from driving with the windows down and moving his loads out in the sun. I can barely recall him without a beer in his hands. My mother probably met him along Route 1. She was pretty enough to waitress at Caruso's Diplomat, a swank cocktail lounge that considered itself to have a touch of Vegas but was really just another pull-in place along the highway, where motels, strip clubs, roadside diners, and blue-and-orange Howard Johnson's restaurants rose from the landscape. In the 1960s, though, Caruso's Diplomat was a destination. John F. Kennedy held a fund-raiser there, and hockey great Bobby Orr later had a party at the Diplomat to celebrate turning twenty-one. But my mother didn't come home with any of the suit-and-tie men who drank dry gin martinis or scotch on the rocks. She came home with Dan.

Dan was a loner, something that may have been an advantage for hauling, but even off the road, he kept mostly to himself. We never went places, except occasionally down to the truck yard, where I climbed up high in the cab and looked around, or rode for a couple of miles on a very short haul with him. I didn't know it then, but he was apparently stealing things from the company, extra supplies and other stuff, and selling them on the side, hot off the truck or from the shop, for cash that I doubt he ever brought home. What time he didn't spend behind the wheel or on the couch he spent with his car, a shiny Mustang that he kept polished and waxed. He would lose himself there, head buried under the hood. He could make the engine purr. Perhaps he should have been a cylinder, a transmission, an inanimate thing.

We stayed in Wakefield for a while, living in that same, small place, which was really half of a house tacked on to someone's old barn. It had a screened-in porch, and about four rooms inside. It didn't face the street, but was tucked back behind another home. You got in by parking the car on a gravel space and walking along a narrow gravel path to the small square of front yard. Inside, I had a bouncy horse that I could ride for hours, rocking on its metal springs, and a little metal chair, where I sat to watch television.

The Christmas after I turned five, I got a cat, probably in my stocking. It was a small, vulnerable orange tiger kitten, which I named Tiger, with a soft coat and spindly legs. I would put toys in its path, let it chase string, watch it as it methodically washed its face or curled up in a tiny ball, its pink nose twitching ever so slightly as it slept. Cats were easy. They were allowed in rental places; they did not need to be walked; they ate very little. They were compact. But my kitten was not compact enough for Dan Sullivan. One evening when he was spread across the couch with his can of beer, the kitten hopped up, and Dan smacked him with his burly arm. He wasn't sharing his sofa. The kitten gave a startled cry and sailed into the air. It landed not catlike, but in a heap on the ground. I picked it up, but it couldn't stand. It mewed, and I ran my hands gently over its fur. My mother took it to the vet the next morning. He said the kitten's leg was broken and it wouldn't ever heal properly in such a little thing. There was no recourse but to put it down. This was the first sign to me that Dan was trouble. It was also the first time I learned about the feeling of hate. I hated this man, the man who had killed my kitten.

Not long after that, we moved east to Revere, the blue-collar town that bordered the ocean and the place where Dan was from. In Revere, we lived far away from the water, in a split-level home that had been broken up into two dwellings. Our apartment was on the bottom floor, with a basement and a garage underneath; the owners lived above us on the second floor. There was a public beach where I could go, if my mother took me, to dig with my pail in the sand; I still recall the special excursion to eat dinner at Kelly's Roast Beef.

But my first clear memory is not of the beach or the old brick hall in the center of town, or Boston being a mere five miles away. It is of Dan, in bed, early one morning. My mother was in the hospital, giving birth to my sister Leeann, and I was supposed to wake Dan up in the morning, either to see his newborn baby or to be there for the birth. I never quite understood which one. The sun was already up when I tiptoed into the bedroom and watched him sprawled on the bed. And then I shook him. Just a small tug, but nothing happened. I tugged harder. I pulled and I prodded, the smell of alcohol stinging my nose. I had never quite understood before what drunk was, but I learned that morning. He was a combination of drunk and hungover, and he would not get up. But I had been told to wake him, and I knew that my mother needed him. For over an hour, nothing worked. I prodded and poked and climbed on the bed, and then finally his eyes opened. But it had taken me a very long time.

He rubbed his face and caught sight of the clock, and the next thing I knew, he balled his hands into fists and began smacking me around. He pounded my head, my back, and plowed into me with those massive knuckles and flat, sandpapery palms until I was shaking and sobbing and snot was pouring out of my nose. My skin stung from where it had been smacked red. I tried to drop my head and pull up my arms, but he was big and strong. If I tried to sit down and make myself small against the floor, he would haul me up by my arm, his fingers squeezing like a vise against my bone, and with his free hand deliver a clean hit again. I had made him late. It was all my fault. "Stupid-ass kid," he yelled on the day his own daughter was being born.

Then he was done. I was still sniffling, and I cringed as he raised his hand. But it was only to cock his finger, not unlike the way other men cock the hammer of a gun. "If ya tell your mother," he barked, "I'll kill ya." I was still sniffling, but I nodded my head. "You hear me? Tell her and I'll kill ya." And I knew right then that he would. When I looked at Dan Sullivan, I already knew that he had killed my kitten. And I knew that he could well kill me. I was six years old and completely alone with him. It was a feeling of fear and of helplessness that I could barely comprehend. From that moment on, I could no longer be a regular child, no longer run down the block or in the door without looking over my shoulder. I had to grow up faster than the other skinned-knee kids around.

We went to the hospital to see my mom, and I never told. If she saw any marks on me, I probably told her that I fell or Dan said that I was horsing around. Or she may have seen nothing. But I had seen those hands. After that morning, I knew that every minute I had to pay attention. Every second, I had to be watching. And I knew that he would be watching me as well.

Now, when I walked around the house, I was always aware. I don't remember the snowfalls in the winter when the sky turned gray and the trees, cars, and ground were buried in white. I don't remember packing snowballs or making snowmen or lying down to make snow angels in the frozen ground. I was watching for Dan. I was listening. And one night it came.

I woke in my bed to the sound of screaming and banging. I leaped from my covers and ran in my pajamas down to the basement area, toward the sound. My mom was screaming and yelling, and crying big choking sobs, and he was hitting her, his fists landing blow after blow. She'd grab at him, push and claw, but he always managed to free one hand and curl back an elbow for a hard swing. The last thing I saw was him balling his fingers and raising his hand. I dived down. His legs were hard and strong, but I grabbed on with both arms and then I opened my mouth and I bit him. I bit him right through his pants, as hard as I could. I was like a pit bull and would not let go. He tasted of soiled Dickies fabric, of coarse male hair and sweaty skin, but I bit down hard, right in the inside of his thigh, and just as I had seen him do, I made a fist and began trying to hit him. I kept one hand locked around his leg while I swung away with the other one, trying to make contact, aiming for whatever was closest, his backside, his groin, or his balls.

He yelled, took a staggering step back, with me and my mouth still locked on his leg, and released my mom. Then he reached down with those massive forearms. He began pounding on my head until my brain rattled like a Jell-O mold turned upside down. But I knew I could not let go. When he hit me, I bit. Hard. Every time he knocked me away, I'd rush at him again, low and fast, like a crazed lion. He was beating the crap out of me, but I knew I couldn't stop, and with every blow I grabbed and held on. I was like the lightning rod, the metal shaft that absorbs the jolt and conducts it to the ground. He hit and I absorbed, trying to channel the power into my teeth or my little balled hand and then back into him.

The house's owners lived in the apartment upstairs. They sometimes watched me when my mother was gone. Thank God they were home that night, and thank God they heard the screaming and banging. They called the cops and the cops came. And it was over. I never even heard the wail of the siren over the wailing in that room that evening.

It took a few more months before Dan Sullivan was gone. But from that night on, I knew that I had to be the man of the house, that I had to be the protector above all other things. In my six-year-old brain, I told myself, "I'm going to save my mom and my sister." That was where it began.

Before Dan left, I was riding in the car with my mother, standing up in the backseat, goofing around. Suddenly she stopped short and I went flying forward, mouth open. My face hit first and some of my front teeth were embedded in the vinyl dashboard of the car. They needed to be pried out, one by one. They were only baby teeth, but they were so mangled that my mother had to take me to a dentist in Wakefield to remove some of them. I remember Dan's voice, hollering, "How are we gonna pay for this? How? You tell me how?" My mother yelled back, and they started in. I did go to the dentist, and my grandparents ultimately paid to fix the teeth that had bitten Dan.

Not too long after that, Dan Sullivan was finally and truly gone. Packed up and vanished. My mother boxed up our things, I rolled mine in a blanket, and we went back to my grandparents on Eastern Avenue. My sister was less than one year old. She has no memory of her father; she did not see him ever again. He never came to visit, never paid child support. We didn't know where he had gone. Decades later, after Leeann was grown, we hired a private investigator, but we still couldn't find him. In 2010, I finally located Dan's brother and called him on a Saturday afternoon. He told me that he had heard of but didn't know anything about Leeann. It was ironic, he said, because his own kids were working on a family genealogy and didn't know that they had a first cousin.

Dan's loss, I thought. His daughter, my sister, is a great, warm, and loving person. It was truly his loss to not know her.

In the next breath, Dan's brother told me that Dan was no longer around. He had died in New Hampshire in 2002, when he was seventy years old. Fell down and hit his head, his brother said, was what happened in the end.

Chapter Three

WHERE THEY TAKE YOU IN

Eastern Avenue, with its assortment of small 1920s Cape Cod–style houses, looked different now that we were back living on it. It looked like home. It was exciting for me to have a house, rather than an apartment, and a yard with leafy trees that rustled with the constant breezes. I settled into the brown, peaked-roof clapboard house with its living room in the front, den or study behind, dining room, tiny kitchen in the back, and four bedrooms upstairs. I have no idea what it must have been like for my mother to go fleeing back to her parents, with little more than the wreckage of two marriages trailing behind her, and with a baby daughter and a nearly seven-year-old son.

"Home is the place where, when you have to go there, they have to take you in," the New England poet Robert Frost wrote around 1915. In 1966, five years after the aged Frost had read at John F. Kennedy's inauguration, my grandparents took their youngest daughter in, again. In his room, my grandfather no longer slept alone. At night I bedded down in the same room, in a second twin bed with a nightstand in between.

I was old enough to explore outside, to visit the neighbors up and down the block, and to make friends with their kids. I already had a bicycle, a Stingray with the high handlebars. I had learned to ride out on the street with my mom watching. I'd pedal, fall, scrape my knee or elbow, and then get up again, until I figured out how to balance and keep going. By now, I was an accomplished rider. I could pick up speed on the hill and feel the icy New England air slice through the buttons of my jacket, burning the back of my throat as it rushed into my lungs. In the cold, I could breathe smoke, crystallized puffs from between my lips like a dragon. In the summer heat, speed was my own private gust of wind. With my legs pumping against the pedals, I could outrace everything, forgetting the time when I had first learned to ride and thought I could race down the hill of the Avon Street sidewalk with no hands. I lost control and flipped over the handlebars, sliding down the road on my chin and splitting it open. Fortunately, a doctor lived on the street, in a white house with a beautiful bay window. My mother took me over, and he stitched me up with nine stitches in his small home office.

Not long after we moved back to Wakefield, my dad showed up a few times again. I don't remember him coming at all during the time we lived with Dan, but perhaps he did. Or perhaps he had waited until there wasn't a man around again, or perhaps my mother kept him away. I just knew that Dad had temporarily returned.

I would wait for him at the front door of my grandparents' house starting sometime on a Saturday morning. If he said 10 a.m., I would be ready at 8. I can remember one time standing at the door with my nose to the glass, making circles in the air with a toy airplane. I spun it through my fingers, letting it soar, roll, and dive-bomb, all the while keeping my nose firmly against the cool of the glass. My plane flew as far away as Chicago in the hours that I spent at that doorway, and my father never came.

When he did come, it was on his own time. He might

pull up around noon or later. Hours on the clock were long and elastic to him. We'd go for a drive, maybe stop at a diner, and it could have been all of an hour before he was angling the car's wheels back against the curb, shifting gears and receding down the block again. I don't remember what we did; I just remember the waiting.

With Delores and his new family—which included my half sister Robyn, and soon thereafter a second son, Bruce—my father lived about thirty-five miles away. He had a place over in Newburyport, along the edge of the Merrimack River before it empties into the Atlantic above Boston. It was a quaint, old saltbox-style Colonial with high ceilings and old pine floors, large windows, a shiny new kitchen, and a comfortable living room. Newburyport was a run-down fishing port then; the houses were old, but cheap. There was a time when you could barely give away real estate there. But my father saw potential in it. It was also part of his sales territory, and a place where he could buy a nice, comfortable home. There were plenty of bedrooms, for him, his wife, and their two children, but no spare room for me. The rare nights when I did stay over, I bunked on the couch with a pillow and a blanket. My father often left me with Robyn and Bruce, under the theory that we were all his kids, and thus siblings, and so of course we would play together and get along. But I didn't see it that way. They lived in this nice house; they went on vacations, to the beach, to Disneyland. I had never even been to Boston; I knew nothing of the Public Garden or Faneuil Hall or Fenway Park and Red Sox games. I only had to glance around to see the incontrovertible evidence of his new life.

As each year passed, photos of Robyn and Bruce and Delores were displayed in frames all over the house. I have a clear memory of wandering around the house, wondering where I fit in, and noticing that my father hadn't put out a single picture of me, not even a school picture or the baby snapshot of me surrounded by his trophies. That was what hurt more than the bedroom. There wasn't one photo. I was, in a sense, airbrushed out of his life. He had a new car, a new family, new kids; I was the afterthought. If he had to introduce me to anyone, he would say, "This is my son, Scott. He lives with his mother."

But my father was charming. He was tall, about six foot five, and handsome with a brilliant, light-up-the-room smile. He could tell a story or crack a joke with the best of them. When he came to get me, he would sound smooth just by coming in the room, and even my grandmother was not completely immune to his charms. She retained a soft spot for him; my grandfather looked at the man in his living room and no doubt remembered the unpaid loan.

My father sold insurance now. I had no idea how he ended up in that business or why he did it in Massachusetts, except that his manner must have made him a natural salesman. He had a way of talking that drew people in. My mother still talks about how he swept her off her feet. During my growing up, he spun stories for me, told me that he had been an Air Force captain, a navigator in fast military planes. His greatest stories concerned basketball. He said that he had played basketball first in Pennsylvania, where he grew up, and then in the Air Force. In between, he said, he had played pro ball in Cincinnati, stepping in for Maurice "Maurie" Stokes, the 1956 NBA Rookie of the Year. In the last regular game of the 1958 season, in March, after a drive to the basket, Stokes hit the floor, smacking his head. Three days later, he fell ill and slipped into a coma, from which he awoke permanently paralyzed. My dad told me that he had played for Stokes for a short time, but long enough so that Topps made a basketball card with his face on it and a bunch of stats, a card for C. Bruce Brown. He never showed me the card, and I never bothered to look for it. I also never bothered to do the math, which would have pointed to a different conclusion. My parents met in the summer of 1957 and married six months later, and in the Air Force, my father was an airman first class, an E-3 enlisted man.

I did know that he flew airplanes, although not in the Air Force, and I never went up in any with him. The few times we were going to fly, we would get ready and then something would happen. The closest I got was when he used to drive over to the nearby airfield and plunk me down with my little half siblings to watch the twin-engine planes do takeoffs and landings. But friends of his say that they have flown with him and that he has a pilot's license. And I know he had trophies from his basketball playing, but where he won them was vague. He never talked about his own family. I know his mother was named Delores, and that she drank and was what people in the 1930s and 1940s liked to call a "loose woman," although in the photo I have of her, she is a delicate-featured woman with wavy dark hair and a taste for intricate jewelry, in this case dangling earrings and a necklace of colored glass or luminous stones.

If you offered me a million dollars, I couldn't tell you anything about my father's father, not even his name. I was fifty years old before I saw a good photo of him—in one fuzzy image, he looks like a barrel-chested man with a hint of a belly wearing a houndstooth blazer over some kind of T-shirt, his hair puffed up in the pompadour style of 1930s or 1940s men. The photo looks as if it was taken in the crowd at a racetrack, and my father's father is not even looking at the camera—his eyes are off following other action. The suggestion was that my dad had been raised by his aunt and uncle, until he got out of that small Pennsylvania town.

He arrived, when he came, in a convertible. This was always the kind of car that he drove, in contrast to the image of an insurance salesman behind the wheel of a staid four-door sedan. My dad is a man who likes the top down. On our drives, when he wasn't spinning stories, we made small talk; we basically talked about nothing. He was not a man for father–son talks; I was more likely to pry some stray fact or truth from my grandfather, in between puffs on the pipe that he loved.

With my dad, I would wait for weeks, wanting him to come, and when he finally did, the reality could never meet the expectations. And truly I didn't know what to expect. My dad later said that my mother would never let him see me, that I was a pawn in their wider war. He would say that she was unpredictable and volatile, and always confrontational, although in a different breath he could remember her beauty and how funny she was when they first met. My mother would retaliate and tell me that my father was a womanizer, someone who couldn't be trusted and who was only out for himself. I don't know. I never knew the truth. Fifty years later, the story is still too loaded from both sides. The only things I know are the things that I can see with my own eyes, or that I can touch. I cannot take anything fully on faith; I require, as the law puts it, third-party verification, and verification from someone I trust.

Yet buried within the layers of stories and embellishments sometimes is the truth. Years later, when I was grown, my father told me that he worked for Jay Leno's father. When Leno's dad was the regional manager for Prudential insurance, he hired my father as one of his salesmen. My dad told me that he used to go over to the Leno house, that he knew Jay and Jay's mom. And that story is true; my father hung around the Leno home enough that the family named their dog Claude, after my father, Claude Bruce Brown. Jay himself told me so when I was a guest on his show in 2010. And I know that my dad probably saw more of Jay Leno as a teenager growing into a young man than he ever did of me.

All my other friends lived in houses with both their parents, in what I thought of as happy, stable families. Up and down my grandparents' block, the neighbors knew that I was Scotty, the boy who barely saw his dad. I would roam from yard to yard, from Eastern up along Court Street, as far as Aborn Avenue and back over to Sweetser Street, popping in, and they in turn acted as if they almost expected me, opening their doors. But in my six-year-old mind, I tried not to be a pest. If a neighbor was raking, I would help. If another was mowing the lawn, I would pick up the leaves. Sometimes neighbors gave me a quarter or a cookie and a glass of milk. Within a few weeks, I knew most of the neighbors: Mr. Spicer, who lived across the street and tinkered with car engines; or June, the older woman who lived alone in the small white house with black shutters at the end of the block just across Court Street and who had me in for "tea," which always involved a small plate of cookies at her kitchen table. She asked me to call her "Aunt June," and I did.

I made friends with David McClellan, one of the McClellan boys who lived three houses down on Eastern; and Susan Norton, who lived next door. No one turned me away, and whatever they thought of my mother and her marriages and her two children, the neighbors kept to themselves and I kept appearing at their doors.

Not long after we moved back to her old home, my mother changed Leeann's name. It was too confusing to have Scott Brown and Leeann Sullivan. It would require correcting people after introductions had been made, because who in 1966 or 1967 would assume that a brother and sister would not share the same last name? The simple fact of our different names would be giving away my mother's life story, a story that she was not particularly keen on explaining. My mother petitioned to have Leeann's name formally changed to Brown, and it was. My disappearing father now had two offspring, and Leeann now had two phantom dads.

After the apartment in Revere, my grandparents' house was like an oasis, down to my grandfather's steady rumble of breath at night when we slept in our shared room, until I graduated to a daybed in the tiny front space, under the eaves. My grandmother kept a supply of crafts on hand to keep me busy, especially ceramics, bits of tile that I made trivets and mosaics from, and Popsicle sticks, which I turned into jewelry boxes and other creations. There was always a crafts table with a project, like making ornaments for the Christmas tree or for Halloween, even in July. My grandfather had already taught me about raking, and as I grew, next came the push mower for the lawn. When the winter cold receded, my grandmother had me tagging along in the garden, weeding, planting tomato seeds in little pots, waiting for them to sprout, and then burying them back into the ground, staking them, and waiting for the red fruit. In the evenings, Grandpa taught me to play canasta, and he, Gram, and I would play for hours in the evenings. I hated to lose, and I still remember him always chuckling when he invariably won.

Gram and Gramps were the source of most fun. When there were birthdays or holidays to be celebrated, Gramps was the one who took us all out to the Hilltop Steak House, sitting on a rise along Route 1. People drove for miles and lined up for hours to eat at Hilltop, which had a giant green cactus perched in front along the road, as incongruous a landmark among the oak trees as the oversize orange T. rex at the miniature golf park a few miles south. At Hilltop, the baked potatoes came out wrapped and steaming in silver foil; as I grew, I would order the filet *and* the lobster pie, eat them both, and always want more. Gramps liked Hilltop, but he liked McDonald's too, and he watched the sign out front underneath the golden arches, checking for when they had sold their next million.

He occasionally took me on larger outings that he deemed suitable for a young boy, driving across town to Pleasure Island—unveiled the year I was born with the hope of being the Disneyland of the East, offering boat rides through Pirate Cove, where passengers searched for the great white whale—or jaunts on the Slanty

Shanty, the Jenney cars, and Old Smoky. A few times, we'd head south to Carver, Massachusetts, where the Edaville Railroad ran, with old-fashioned miniature trains and a small petting zoo. He particularly loved the trains, the sound of their tinny whistles straining and puffing, the rhythmic clatter as their wheels clacked along the rails. As an engineer, he preferred the orderly mechanics and precision of trains, arriving on time, departing as scheduled. My grandmother liked the animals. Her family had been farmers—some of her siblings still worked the land, and I distinctly remember visiting rocky New Hampshire farms and gazing at the barns and animals.

Even with the activity around Eastern Avenue, I could barely sit still, and after living under the shadow of Dan Sullivan's fist, it was tempting to break the rules, and I did. I would take off without telling my grandmother where I was going or return home late for lunch or dinner. After one such infraction, my grandmother began chasing me with a rolling pin in her hand, probably to spank me, but I eluded her second by second, diving under the table, and then just as her hand began reaching toward me, I took off again, taking cover behind the breakfront, racing off to the couch and then up the stairs, always maneuvering just out of reach. All over the house we went, my exasperated grandmother trailing after me, like a live-action version of Tom and Jerry or Bugs Bunny and Road Runner cartoons. She soon learned that when I got into too much mischief, the most effective punishment would be to force me to sit still for a half hour or longer on a red vinyl chair in the kitchen, facing the clock. There was nothing more excruciating than watching the minute hand move like a glacier from one side of the face to another. The moment my time was up I would be out the door like a shot, racing down the block, banging on the door for David or Susie to come out, and calling for them at the top of my lungs, heedless of whether it was early or late, lunchtime or dinnertime. Looking back, being active was the best medicine for me.

It's ironic that my grandmother was my chief disciplinarian, however much she was a stickler for rules, because it was my grandfather who ruled their home. Gram was much more of a free spirit. Leeann later told me that when Gram was a teenager, growing up on her parents' isolated New Hampshire farm, she would drag an old mattress down from the house, across the lawn, and toward the river, plunk it down on the grass, and sunbathe in nothing more than her birthday suit. She never told me that particular story when I was a child, but she did often say, when it was just the two of us, how she wished that she had never given up teaching, which she loved. But even with her children grown, returning to the classroom was completely out of the question. My grandfather went so far as to discourage her from wearing lipstick; he frowned on any type of what he called "face painting." He was a man who expected to have his dinner waiting for him on the table at six o'clock when he came home.

Late each afternoon, when her pots were already on the stove or the oven was on, Gram would gather Leeann and me up and put us in the car to head down to the station to pick up Gramps. The Eastern Avenue house was close to the train line, and all day the whistles sounded off, their high-pitched toots like a clock chime to Gram, counting down the hours from when Gramps left the house until he came back home. Philip Rugg was a thrifty, tradition-bound man, but I also learned that he kept a thick file folder of all my accomplishments—I have my middle name in honor of him—and he found odd jobs for me to do around the house to earn extra money. He was constantly helping out my mom. And, of course, they were also looking after Leeann.

I loved him and I loved my grandmother, and they loved each other, but any emotion was restrained. What had begun as love had long since fallen into a more tight-lipped respect, affection, and lifelong dedication. They were married for over sixty years. Endurance was more the state of their marriage and resilience its overwhelming quality. I can still hear him saying, "Oh, Bertha, Bertha, Bertha," blowing out the words along with his pipe smoke, shaking his head, and then pausing to inhale again when my grandmother was going on about something or other. For her part, Gram would pretend not to hear him. Instead, she would occasionally escape to the pantry closet, where she kept her cooking sherry, and pour a glassful or down a quick nip straight from the bottle to take the edge off the small house when they were both at home. In the summers, she escaped entirely, with me in tow. We would head north to Rye Beach in New Hampshire and rent at Hoyt's Lodges, little, narrow cottages painted white with blue awnings, where we stayed for as long as three to four weeks at a time. Grandpa drove up once in a while for dinner, but most days it was just the two of us alone.

Each Hoyt's cabin was like an efficiency unit with a single bedroom—just large enough for a bed and chair and a chest of drawers tucked in a wall niche—a bathroom, and a small kitchenette, with an electric stove and a round, white mini Frigidaire. The walls were paneled in knotty pine, and just to poke your head in from the small screened-in porch was to smell that slightly musty, damp wood smell of the New England coast, bottled up inside the room. In the mornings I would eat Cheerios, or some goodies that Gram had cooked, or sometimes a doughnut, warm and sticky with glaze, and drink orange juice, probably from concentrate frozen in a can. Then Gram would pack a picnic lunch and she and I would head down to the beach, which in this part of Rye is really more of a rocky coastline, like Maine. I spent hours climbing and scrambling along the dark, massive rocks and exploring the caves and caverns and swirling tidal pools that dotted the shoreline, where small shells and skeletons came to rest after they had been tossed about by the salt water and bleached by the sun. I also collected jars and jars of sea glass, weathered browns and greens and the rare deep blues, a tradition that I would one day pass on to my own children.

I could climb up the crevices and find the flat indentations for my feet amid the jagged edges and slick corners washed by the tides. I would go as far out as I possibly could without falling into the dark, chill water below. Of course, the day was not complete if I did not return with my pockets stuffed with crabs, starfish, and sea glass, and me halfway soaked from losing my footing and slipping in. Gram never went out on the distant rocks. She sat on the shore, which was simply a swath of smaller rocks and pebbles—to get actual sand, we had to go south along Rye Beach or down to Hampton—and read, watching me recede and then return along the rock line. Occasionally, she would explore too, dragging long threads of seaweed behind her, picking out sand dollars and seashells.

In later years, when they spent more time in New Hampshire and Gramps also came, he would wait in the car, parked in the little asphalt turn-in, and sit for hours while Gram, Leeann, and I cavorted along the seawall and scavenged for whatever treasures we could find.

When I wasn't climbing, she tried to get me to color or do crafts, or we might take a trip down to the wharf or the pier to watch the lobster boats coming in or going back out to sea, their replacement metal traps stacked and tied to hopeful, fading buoys that would bob in the waves. She took me on long, rambling walks through a nature preserve and out to Odiorne Point. We ended at a wooden bridge, where I bounced rocks and watched them skip in the water below. At night, we watched television or listened to the radio. We might sit around and talk, but Gram always put me to bed early, knowing I would be up not long after the sun.

We went to Rye Beach for about five summers, from the time I was five until after I turned nine. We went there from the time of Dan Sullivan, from the era of tiny apartments and the two interludes of houses, houses where I lived but which were not my own. My dad might come on a Saturday, or he might not. My address might change every six months, but for half a decade, those weeks with my grandmother were fixed and orderly, like the moon traveling around the earth and pulling the ocean tides along.

Then, with no warning, life at Eastern Avenue came to a screeching end. Midway through my second-grade year, my mother announced that the three of us were moving to Malden. She was marrying, again.

Chapter Four

MALDEN

My mother met Al Di Santo in Wakefield. He was a bartender at the Colonial, a large restaurant and function facility adjacent to a golf course. But he lived two towns away down Route 1 in Malden, Massachusetts, a town that dates back to the Puritans, who noted their discovery of "an uncouth wilderness," and bought the land from "a remnant" of the once powerful tribe of Pawtucket Indians, according to the meticulous 1880 *History of Middlesex County* by Samuel Adams Drake. Like many budding Massachusetts towns, it tried one or two names, and was known for a while as "Mystic Side," for its proximity to the Mystic River, before a group of local settlers petitioned to have its name changed to Maldon, named for Maldon, England. Two centuries later, someone inexplicably altered the spelling to Malden. In the 1690s, during the Salem witch craze, Malden imprisoned two local women for practicing witchcraft. In 1742, the town census valued both an "oald negroman" and a cow at ten pounds each. In April of 1775, Malden's seventy-five-man militia was called to arms for the Battle of Lexington and may well have captured a British supply line. Over time, Malden evolved into a factory town, making nails, dyeing silk, and manufacturing shoes and pieces of block tin. It would become home to the Boston Rubber Shoe Company, whose treasurer was a man named Elisha S. Converse. But I never thought of the Converse sneakers that dotted the basketball courts as having some connection to my new home. What I knew was that across Route 1, Malden bordered the town of Revere.

Al was older, probably about forty-five or maybe just beyond; my mom was barely thirty. He had been married before and had two children, both of whom were much older than Leeann and me, and so much removed from his own life that I don't remember meeting them. Al himself seemed ancient. He was literally

Old World, probably first-generation American. I met his white-haired parents a few times. They spoke only bits of broken English, and most of their talking was done in Italian.

I don't have any memory of meeting Al more than once before we moved in, although maybe I did. Or maybe my mother kept Leeann and me tucked away, hoping that once they were married, Al would adapt and adjust. But Al did neither. He looked at us, Leeann a toddler and me a rambunctious seven-year-old, and was overwhelmed. He didn't want two young children touching his stuff, bringing disorder to his home. He wanted a life without kids, and he didn't get one, unless my mother was able to drop us at my grandparents' house. When we moved to Malden, C. Bruce Brown went into a kind of exile, or was dispatched from my life by my mother. As in Revere, as with Dan Sullivan, he did not come around much at all. Like instant soup, Al had an instant family, and the temperature was always set slightly below boiling. He was frustrated, and he spent his time grumbling.

His house was a small split-level on a cul-de-sac, Como Street, named for an Italian family who lived across the street, worked in construction, and probably built most of the homes. The Comos had a seven-year-old boy, and Bobby became one of my best friends.

From the beginning, I knew that Al's house was his house, and his alone. I walked from room to room as if I were tiptoeing around a museum, feeling Al's eyes upon me even if he wasn't home. But he was home a lot. He worked bartenders' hours, and was gone in the evenings, to pour drinks for banquet-goers or restaurant diners. Unlike Dan Sullivan, Al was a small man, about five foot eight and thin, sort of like Sammy Davis, Jr., looking insubstantial but somehow graceful even in his black bartender's suit. He had thick silvery-black hair, wore wire-rimmed glasses, and smoked all the time, until it was hard to tell where his fingertips ended and the cigarette nub began.

While I had been grateful to get out of Revere, this time I had not wanted to move. I did not want to leave Eastern Avenue, to leave my grandparents and the neighbors' open doors, my school, my little group of friends. But my mother plucked us up with our few possessions and made off for Malden. We didn't live in the local downtown—we were high above, on one of the cliffs. Behind Al Di Santo's house was a sheer rock wall that I scrambled up and down, rocks that were probably the legacy of a million years of glaciers, advancing and retreating across the Massachusetts landscape. As the glaciers drifted down from eastern Canada, as much as ten thousand feet of ice accumulated, compacting the lush ground where dinosaurs and other prehistoric reptiles had once roamed. When the glaciers melted and receded, much of the fragile soil went with them, while the bedrock remained. The hill on which Al Di Santo's home stood was the legacy of millennia of geological struggle.

Al's foundation itself had been built into the rocks, and rising up from the floor of his basement was a giant cliff stone, tall enough that it almost touched the ceiling. I liked going down to the basement, and imagining the stone as a fort, I would shimmy up it and slip myself into the narrow space between the top of the rock and the ceiling. But Al didn't like Leeann or me down there, where he had boxes of his stuff, tools, and other private things.

Without my grandmother's projects table, I tried to create my own amusements. And whenever possible, I tried to do them outside, beyond the confines of Al's untouchable furniture and walls. One afternoon in the backyard, not that long after we had moved in, I tied a two-by-four to a rope, swung it over my head like a lasso, and sent it sailing in the general direction of my sister. It hit the mark, slamming right into her head, bloodying her nose, and sending her sprawling to the ground. My mom heard Leeann screaming and raced outside. When she asked me what had happened,

I probably said something stupid like the wood had fallen from the sky. Somehow she didn't place much stock in my explanation, especially when she saw the rope, the two-by-four piece, and Leeann. Naturally, she immediately began screaming and waving her arms at me. Everyone was yelling and screaming. I was in tears. I had never wanted to hurt my sister; it was a backyard prank that had gone horribly wrong. But my mother would never allow me to forget that afternoon. She untied the two-by-four and kept it on the top of the fridge or some other storage spot in the house. When I misbehaved, she would haul it out and whack me with it. For years, she kept that vicious block of wood, until one day when I was much older and had finally grown too strong. Then, with my teenage hands, I grabbed it from her and threw it away, daring her to go get it. She never did, and the dreaded wood was gone.

I don't remember my mother fighting with Al the way she fought with Dan. He wasn't one to hit and he was never cruel. He actually was a good listener, a skill honed during all those nights of tending bar when people poured out their troubles to him over a scotch or a beer. He could have a temper, but I was never scared of him. He was not a violent man who raised his fists, and I grew immune to the strings of curses when he on occasion spewed them. Words, I felt, were nothing.

My mother and Al were both working, often at night. Many evenings, my mom would drop me and Leeann off with Gram and Gramps; we would play at their house and then fall asleep there in our pajamas. After her shift was over, my mom collected us, asleep, and drove us back down Route 1 to Al's place in Malden, where it was easy to hear the rumble of the highway whenever the windows were open.

I was always in trouble, at home, in school. I seemed to gravitate to it, as if it had tentacles that it could unfurl and draw me in—as when I decided to flick matches at the edge of the woods in a neighbor's backyard on Como Street because I was bored and looking for something to do. I did not have a good history with matches. When we lived on Eastern Avenue, I had tried to build a campfire inside my grandparents' garage. First, I gathered sticks and dry leaves from the corners of the yard; then I brought them into the garage. I stacked up my fire makings, struck a match, and touched it to the dry kindling. I was expecting a little fire, but instead the flames took off, almost taking the garage with them.

That should have cured my interest in fire, but it didn't. One day at Al's house, feeling bored, I grabbed a matchbook and wandered outside, flicking the matches against the striker. I walked into a friend's backyard, still flicking the heads and watching them catch fire and sail through the air. They ignited like little rockets, gliding on the wind until they were consumed. But one didn't—it reached the twigs and leaves at the edge of the woods in the back of one of the houses on Como Street. They were dry and waiting to ignite. I heard the crackle of fire and for a few seconds stared in disbelief as the twigs and leaves caught, the fire crumpling them up and sucking them in. They popped and sparked as they burned and the sparks carried, until it wasn't just one little patch of leaves but an entire section of ground. Panicked, I raced over to stamp out the flames, but they were moving fast, fanning out across other dry leafy patches, and the heat was rising, melting the soles of my sneakers as I jumped up and down.

Water, I thought, and ran to get a hose, but it was dry and I couldn't get the outside tap on. Buckets. I grabbed two and filled them at another house, but the fire had already started snaking back into the woods. The underbrush caught and then the trees, which made loud crackling sounds as they glowed and turned orange. I had dampened one foot of ground, but that meant nothing. The flames had taken on a life of their own. Frantic, I pounded on a neighbor's door and yelled to call the fire department. The red engines roared up the steep hill, grim-faced firemen in full gear dragging their bulky hoses across the ground and spraying down everything, until there was only black char, and curls of gray smoke were hissing up from the ground. It took about an hour to put the fire out. The police came too and gave me a warning about matches, telling me never to flick them again. "You started a forest fire!" my mother screamed at me. I don't know what Al said to my mother about having a kid under his roof who would walk out with a pack of matches, but of course with all the cigarettes, there were always plenty of packs lying around. It didn't matter. My mother was hysterical enough for both of them, and I got smacked with the two-by-four for that one.

The next morning, the air smelled like soot, there were burn marks at least halfway up the tree trunks, and the ground looked as though someone had come and shaved it too far down. I wasn't allowed out of the house.

But I wasn't entirely chastened, and I was not terrified of fire. One day, while we were still living in Malden, my mother caught me sneaking a cigarette, lighting the tip and trying to inhale. She didn't give me a lecture on smoking. She yelled for a few minutes and then she made me eat the cigarette. All of it. The white paper liner, the cured tobacco leaves, the filter that I had just moments before been holding between my teeth, and the singed end. She stood over me as I chewed and swallowed, choking down the last bite. "That will teach you to smoke," she said. That and throwing up the chewed cigarette pieces afterward, the tobacco-tinged stomach acid burning my throat. Even as my mom and Al stubbed out their butts in ashtrays or tossed them into the bushes, I never went near another one, not even when I was in college or grown.

I was the kid who couldn't sit still. As I had done on Eastern Avenue, I roamed. I made my way down to Broadway Street. There was a tile store there, and I used to root in the Dumpster for bits and pieces of broken tile, gather them up, and carry them home to make things to give as presents for holidays or for special occasions. I met a black family who lived on the main road, just beyond our backyard, and made friends with their son, and for a while the other kids in school would chase me home, yelling a few choice slurs. I had no idea that places like Newark and Detroit were about to ignite in race riots, with buildings going up in flames, or that Boston would soon face bitter fights over school busing and racial integration. My friend was a kid who would play with me, and I could play with him. When I wasn't with him, I hung out with Bobby Como across the street. We watched *The Lone Ranger* and *Speed*

Racer, rode our bikes, and played army and hide-and-seek across the backyards. I knew some of the older kids on our tiny block, including Jimmy and Henry McGowan. The McGowans lived in a small, white house on the curve of the cul-de-sac. Unlike the rest of the houses, which were bunched up against the street and each other, theirs had a real yard and a bit more land. Millie was their mom, and Fred, their dad, was blind. Jimmy McGowan was about eighteen or nineteen when I first came to live on Como Street, and I thought he was the best, the coolest guy in the whole world. He was happy-go-lucky and always laughing and smiling. Then he got a buzz cut and went off to Vietnam. When he came back, he was a totally different person. He didn't smile. Before then, the concept of war had never registered with me, but now I began to think about war and conflict and soldiers fighting and I wondered what happened. I never got a chance to ask him, and if I had, I doubt he would have said more than a few words, and most of those might have been incomprehensible to an eight-year-old.

He had, I later learned, been an explosives-demolition expert. I was told that he had to blow up the bombs and hidden mines that ripped other soldiers to shreds in the camps and jungles of Vietnam. Henry, his brother, fought in the war too.

I had an absentee father and a stepfather who found me a nuisance, so the closest substitute was the older kids in the neighborhood. They were cool kids who seemed to have no rules or curfews and who could be out on their own. Not long after I moved to Malden, I started hanging around with one guy, a thirteen-year-old teenager who lived nearby in a standard postwar Cape house. Around the neighborhood, he and I often used to cut through a small patch of woods, the Dexter Road woods, which sat high atop our cliffside hill in Malden, two blocks above Como Street. It was the way I sometimes walked coming home from school. At one time, woods must have covered the entire crest, and even with years of clearing and building, this portion had remained intact. They were a partial continuation of the same woods that, flicking my matches, I would set a corner of aflame.

In the spring and summer, the Dexter Road woods were thick and overgrown, a tangle of vines and low branches springing from ungroomed trees, but they were passable in the fall, winter, and early spring, before everything leafed out in full bloom. Sometimes I would race through them, my feet crunching over the old, dried-up leaves, but mostly I wove into and out of tree trunks, inhaling the ground's mossy, musty smell and the scent of decaying wood. To a seven-and-a-half-year-old boy, it was exciting to navigate, to roam where there were no manicured grass yards, nothing to worry about stepping on, no rules.

I was in those woods late, but not that late, on a winter afternoon when my teenage friend appeared. He looked at me and said, "Hey, let's hit the path," the path that snaked through the woods. We walked a little way down the worn trail and then he turned. He grabbed me and smacked me and showed me the knife in his hand. Initially, I did not realize what was going on. I thought he was my friend. Then, I began quaking. My entire body began trembling with fear and confusion. As he held the knife, he reached for his pants. At first, I thought he was going to take a leak. Then he undid his belt, and soon his pants were down. He had a look in his eye that I had never seen before.

He gripped me in one hand, held his knife in the other, and he pointed and told me to put my mouth on him. I had no idea why he wanted me to do it or what he wanted me to do. I was totally shocked, completely uncomprehending, and helpless in the middle of the path. I kept hoping that I would hear the crunch of another footstep, that someone would come down the path to help.

But it was just the two of us, hidden deep inside

those woods. He flashed the knife and told me to get on my knees as he casually stretched himself out on the ground, with the ease of someone who had done this before. I reluctantly dropped down, feeling the chill of the decaying leaves and dirt seep against my pants. It was cold, and we were totally alone. I could have screamed at the top of my lungs and the sound would have been lost to the swaying of trees and the wind.

He settled on his back, with the beginning of a grin on his face, still clutching the knife, and yelled at me, thrusting his hips up in my direction. I lowered my gaze and took a last, frantic look around. I saw a rock. Not a small one, but a good, medium, fist-size rock with some heft to it. While he closed his eyes in anticipation, I shifted for a second and wrapped the rock in my palm. As I started to bend my head, he kept holding his knife, but he briefly let go of me. His free hand was splayed back and he was lying down, waiting, with his stupid grin splitting his face. I knew that I had one chance, just one. As he closed his eyes, I raised the rock high over my head, drove it down into his face and head, and took off. I heard him howl in pain, but I never looked back, I just kept running and running, twisting in between the trees, hearing my feet smack against the old leaves and the hard ground. I ran until I got home and went straight to my room. My chest was heaving, my heart racing. I didn't know exactly what had happened, and I told no one. I was too sick to eat dinner.

That night, the doorbell rang. My mom wasn't home, so Al answered. It was my attacker and another older kid, a friend or brother, at the door. The boy had a big red gash on his forehead where I had pounded the rock into him. They asked Al if I could come out. Looking at the two of them under the glow of the porch light, Al sensed that something was very wrong, but he called me to come over. I didn't say a word; I can't imagine how I looked. I was scared, even terrified, but I knew that I could not show it. I knew they were both there to exact revenge. Instead, I just looked at that teenage boy. I stared at him and refused to turn my eyes away. I wanted him to be as afraid of me as I was of him, to think that I would tell my mom or even tell Al what he had tried to do to me. The kid must have seen something in my eyes that made him believe that I was going to talk. I think he sensed, standing there on our doorstep, that there would be no revenge that night. And I think the fact that I never backed down from him ensured that he never tried to approach me again, as he had done in the woods.

But things did not end there. This kid had a lot of friends, and he began making up lies and bad-mouthing me to them. Every day, they would chase me home from school, the Maplewood Elementary School, a low, tan brick building with rectangular windows over on Laurel Street. I had no choice but to outrun them, even though it was over a mile and a half home, much of it uphill. They were bigger and stronger; they could move with a quick, ground-covering stride. If they caught me, I knew what would happen. To this day, I can still see the flash of that knife blade in the woods and the thirteen-year-old boy with his pants down. I was a fast kid, but I pushed myself to be faster. I ran down Laurel and turned left at Salem, followed it along to Broadway, where the street was busy and traffic whizzed by. I stayed on Broadway until it became Temple and then ran up Elwell Street, which was an almost vertical incline. I was sucking wind; my heart was in my throat. And every few yards or so, I turned my head to look back. I looked around every corner and glanced behind me, to catch sight of them coming. Sometimes, I cut through the black family's yard and scrambled up the rock face behind Al's house, hauling my books and myself into his backyard.

Those boys should have gotten me, but they never did. I became a runner then, and every afternoon became a race that I had to win.

Chapter Five

COUSINS

One day we were living in Al Di Santo's house, and the next day we weren't. When third grade ended, my mother moved out of Malden and back to Wakefield. We did not go back to my grandparents, but to an apartment on Albion Street. Two Hundred Ninety-three Albion Street had been an old New England Victorian, with a few bits of gingerbread trim above the covered porch and an octagonal add-on at the sides—the type of house that might have been conjured by Charles Addams or later Stephen King. Farther away from downtown, the train whistle was only a faint sound. The place was a carved-up hodgepodge of rooms, up a long flight of stairs from the busy road. Our apartment was three rooms, a sitting place in front, a bedroom with a fake fireplace, and a kitchen at the rear. I slept on a cot in the bedroom, pressed up against the false fireplace while my mother slept with Leeann in the bed. I spent most of my time outside on the porch, waiting for something to happen.

Al showed up on our doorstep a few times and tried to reconcile, begging my mother to come back. He apologized and promised to be more accommodating. But my mother was done. After the last time that Al left, I never saw him again except for one quick, curious visit when I dropped by his house completely unannounced after I was grown. I needed to see if what I remembered was true, if the way I remembered things there, and even the neighborhood, had stayed the same. He lived in that house in Malden until he died of cancer, tended to by a nephew. His own children, in California and Hawaii, were too far away to come home for long. He wasn't, in retrospect, a bad guy. He was not violent, and despite the fighting and frustrations, he was generally nice to my mom. But Judy Di Santo, formerly Sullivan, formerly Brown, née Rugg, had decided to move on.

I was moving on too, though I didn't know it then, when the meager boxes were unpacked and I unrolled a scratchy wool blanket where I had bundled up my things. That summer, after a few weeks of a Christian sleep-away camp on Cape Cod, which my grandparents paid for, and a visit to Rye Beach with Gram, my mother packed my suitcase and announced that I was going to visit my cousins. My mother's older sister, Nancy, lived in Wakefield. She had married not too long before my mom. But they were not close. They had always been very competitive. Early on, my aunt assumed the mantle of the older, responsible sister, a bit frumpy and staid, while my mom was the young, pretty one. Even well into adulthood, their roles and images remained. I have a hard time imagining them ever sitting up at night, trading secrets in the small bedrooms upstairs on Eastern Avenue. And there was the fact that my grandparents seemed to prefer my mom. Partly it was that she was far more needy, and not just for money. But even with her three marriages and her disconnected life, there was something about her that they responded to. It was not the same with Aunt Nancy.

Of course, as a nine-year-old boy, I didn't know any of this going in.

My aunt Nancy and my uncle Alban had two children: Kenny, who was older; and Wendy, who was about my age. Wendy and I never really liked each other; it was a kind of wary coexistence, but that didn't matter quite so much because I slept in Kenny's room. He had the left side of the room and I had the right. My bed was pushed up under the eaves and I stored my suitcase underneath. Our window overlooked the roof of the garage. I had a dresser, but I don't even think I filled it. I didn't have too many things. I stayed for a few nights, expecting my mother to come and get me, to take me home. But she never came. It was probably my aunt who told me that I was now going to be living with my cousins, in their tiny Cape-style house with three bedrooms and two baths on Redfield Road. It was a foursquare house, with a living room in front and dining room on the other side, a kitchen behind the dining room, and the master bedroom on the first floor in the back, instead of a den. Upstairs, there was space for only two rooms, Wendy's and Kenny's—one side of which was now mine. Busy Route 128 ran above us at the end of their street, and I could hear the rush and rumble of cars even with the windows closed. I went to their school, Walton Elementary, and every morning, I rode my bike up the steep, winding hill to the fourth grade.

To this day, I don't know why it happened. I don't

know why my mother sent me to live with them, by myself, without her and Leeann. I don't know if she decided that the Albion Street apartment was too small for the three of us. I don't even know for sure where she was living the entire time that I was on Redfield Road. I don't know if it was too much to have someone care for Leeann and me, if Leeann went to my grandparents, or what happened. It's been over forty years, and still, I have no explanation.

I do know that my mother paid my aunt and uncle to keep me in their home, like a boarder or a ward. They frequently took the opportunity to remind me of just that fact. At mealtime, when I wanted seconds, my uncle or aunt would say, "Your mother doesn't give us enough to feed you." For my part, I would later retort, "I should be able to have seconds, you're getting paid." And I remember hearing them discussing the payments and the money, although I never knew where the money came from. Perhaps my mother earned it; perhaps my grandparents gave some of it to her to give to them. But I knew that I was a transaction. I always felt like a poor cousin who came to visit. Never was that more apparent than at dinner.

We ate our meals around the table in their kitchen, and I would always be served last. My aunt would make a plate of hamburgers, and I would be starving. The plate would go around to the four of them, and then come to me. I got one burger, and then would have to wait while each of them had seconds. It did not matter if I ate fast, cramming the food into my mouth, or slowly, chewing each bite fifty or one hundred times. Very rarely was there a second helping for me; my aunt usually made only nine, and of course there were five of us gathered around. I still remember the sight of those hamburgers, juicy ground chuck soaking ever so slightly into the soft white buns, and how much I longed to reach over for another one. I knew too, from the living room talk, that they did not lose money on me while I lived with them.

But in retrospect, I have to be grateful to them, because when no one else would, they did take me in.

My mother did not come around much when I lived with my cousins. She did not disappear for stretches as completely as my father, whose face was so diffuse in my brain that I sometimes struggled to conjure the details of him, the color of his eyes, his nose, what he looked like beyond his blinding smile and imposing size. But my mother had also gone missing. She might take me out on a Sunday and listen to me complain about how miserable I was, about how I wanted to come home. Mostly, though, she stayed away, as if to come to Redfield Road were trespassing.

I don't really remember if my father came at all. But he must have shown up sometime, perhaps that summer, in between Al's house and the move to Kenny's room, because I had learned to play basketball, and I didn't learn that from my cousins. My uncle had put up a hoop in the driveway, and on the black tar, Kenny, some neighborhood kids, and I went at it all the time. I was always competitive, and he was my closest competition. I would beat him at hoop, stealing the ball, rising up on my toes to take the shot. I had started to grow, rocketing up past five feet, which was gigantic for a nine-year-old back then. My cousins were smaller, red-haired and freckled, and although Kenny was a good player, I had inherited my dad's dark hair and athletic frame. I was lean and I could move, angling my body in different directions. And I was fast, fast from all those afternoons racing home in Malden.

My cousins and I competed not just in basketball, but in all parts of life. In the winter, my uncle built a hockey rink outside in their backyard, laying out two-by-fours, spreading plastic sheeting, and then filling it with water and waiting for it to freeze in the cold and sullen winter evenings. Wendy skated, and we played hockey, facing off with our sticks on this homemade, frigid pond. In her living room on Eastern Avenue, my grandmother knit us thick wool socks to wear inside our skates and scarves to wrap about our necks.

Kenny was a Cub Scout, and my uncle enrolled me as well. My aunt was the den mother, and my uncle was the pack leader, overseeing a bunch of boys who gathered in his living room to listen to his instructions. We spent our weeks in meetings earning our badges and building race cars for the Pinewood Derby.

In addition to badges, the Cub Scouts gave out ribbons. There were red and blue ribbons for winning first place in races and other competitions. I wanted those ribbons. I collected them, and when I won blue, Kenny and our other friends sometimes had to settle for second and red. I might be an intruder, an interloper in their family—with its kitchen table dinners and a long list of strict household rules for eating, for keeping your room clean, for the exact minute when you had to be home—but in the driveway underneath the hoop or on the playing field, I was growing into my own man. Even though I wasn't too sure what a man was or should be.

Like my grandfather, my uncle went to an office each day. He also worked in some kind of technology industry or as an engineer. He liked to pontificate and on countless topics was content to suggest that he knew everything. I quickly gave up trying to have a real conversation with him. He did love to fix things. The basement of the house had been turned into an orderly workshop with every conceivable tool: jigsaws and table saws and circular saws, drills, hammers, and screwdrivers ranked by size. He kept the screws and the nails also organized by size in old coffee or vegetable cans. Everything was laid out with the precision of a surgeon's instruments. My uncle was good at fixing things—the entire family was—and he seemed particularly happy whenever he was wielding a tool in his hand.

I got out of the house as much as I could, within the rules. I would head over to the old water department building at the end of the short block, to a concrete retaining wall that held up one side of the highway. I'd take a piece of chalk, draw a catcher's mitt on the concrete, and throw the ball against it and catch each pitch on the bounce back, straight into my glove. I could time the thud of the ball, its sting into the leather, to the rumble of engines passing and wheels thundering over the asphalt. It was a rhythm of catch and release, arcing it back into the air. I counted how many times I could hit the target, in the center dead on. Whatever number I reached, I wanted more. If I could hit it a thousand times without missing, I could leave Redfield. If I hit one thousand, I would up the number to five thousand. I could play that game of fooling myself: if that, then this. I could do it for hours, confident that Aunt Nancy or Uncle Alban wouldn't come looking for me. They were no doubt as happy to have me gone as I was.

Redfield Road was a neighborhood full of kids, the second wave of families who had moved into these postwar Cape houses, living stable, suburban lives. I never thought of how it must have seemed to my aunt and uncle to have to introduce me to their neighbors, and the behind-the-back glances that must have passed whenever they explained that "Scott is staying here with us for a while." Wendy would remind me of it to my face. She loved to get under my skin by saying, "Your mother isn't even here. She didn't want you." Or sometimes she'd ask, "Where's your father?" each word spoken in a singsong, slightly mocking tone.

I spent a lot of time at school as well in the afternoons and on weekends, flicking baseball cards outside along the back wall—the way I had flicked matches, but with far less bodily risk. We had a game where we used to ride our bikes up to the school and flick the cards against the brick. Each player would stand a certain distance back and snap the card. It required the perfect angle of finger and thumb to send the card spiraling through the air so that it would arc down and land just where the cement met the wall. The person with the card closest to the wall would win. I could get my card to land on an angle just leaning up against the wall, one edge on the concrete, another on the brick, and then I would collect an entire hand of Topps baseball cards.

As the months wore on, I stopped simply roaming through the neighborhood and began running away. The only destination I knew in Wakefield was my grandparents' house, which was a few miles away, past the 128 underpass, with the highway roaring in my ears and the tunnel ceiling vibrating overhead. Then I would head out along Elm Street and over the railroad tracks, around the bottom tip of Lake Quannapowitt and up the slow incline of Salem Street, before I could turn left at the leafy end of Eastern. I'd ride over there on my bike, thinking that perhaps if I kept showing up at their door eventually they would take me in. Sometimes I'd stay for dinner, but in the end I would have to return to my cousins' home. For that whole year, though, my grandparents were very loving. They talked to me and took an interest in me in a way that they never quite did with my cousins. To my aunt, of course, it must have looked like déjà vu, my grandparents choosing me over her children, just as she thought they had chosen my mom over her. But my grandparents also knew that I was more of a lost soul, drifting from place to place and house to house.

It was my grandmother who finally collected my Cub Scout ribbons and other awards and pasted them into a scrapbook, who celebrated my accomplishments when no one else seemed to be paying very close attention. On the first page she wrote in large, even, schoolteacher print, "Record of Achievements! Scott P. Brown."

I have that scrapbook to this day. For years, when I've needed to find some balance in my life, I get out the scrapbook and look at Gram's proud, hopeful writing and the careful way she filled the thick pages. The leather cover is separating, some of the cardboard backing is visible underneath, but those pages remain a home for so much of my life. They are my memory chest, even now.

Not long after the school year was over, my stay at Redfield Road came to an end, although I did not know how or why. I still don't to this day. My mother simply drove up one afternoon and I put my suitcase in her car. A few other odds and ends like Cub Scout trophies I wrapped in a blanket and dumped in the back as I got in. I don't know if I ever said thank you to my aunt and uncle, or if I so much as glanced back down the block as my mother hit the accelerator and the car sped on.

After nearly four years of being a guest in other

people's houses—my grandparents', Al's, and my cousins'—my mother, my sister, and I had graduated to living in rentals, floating among leases in Wakefield. Each time we moved, my mother doused the air with scented spray fresheners to get rid of the smell of the last tenants. Most of the rentals had two bedrooms, and I usually slept in the same room as Leeann. I was ten and she was four, but I had missed her that year when I had lived with my cousins. I liked the feel of us under one roof again, even if the places where we lived were more like dumps. "Shit holes," my mother sometimes called them.

My father had started coming around again, in bits and pieces. Mostly, he took me to a diner for a quick meal or a drive, stopping to check in with business associates or friends, or he'd take me out to Newburyport, where he unceremoniously stuck me with my half siblings, expecting us all to be immediate friends. I detested those trips, but I still made them. And I took to waiting for my father's irregular visits on the stoop in front of whatever place we happened to be in.

Whenever my father came inside to collect me, my mother would be ready, claws bared, to spring at him. "Why are you late?" "You aren't dependable. You're never on time." She was furious at him for missing child support checks. She was supposed to get $25 a week or so. He paid when he felt like it, which usually wasn't often. He had a new house and plenty for his new family, while we, my mother would point out, were living in shit holes. She would scream; he would counter. Every pickup dissolved into a shouting match, until I learned that the easiest thing was to stay outside on the stoop, for however long it took, rather than take a chance that he might actually come in.

After Albion Street, my mother moved around the corner to 202 Broadway, a brown, unadorned New England clapboard. We had the second floor. The house was an architectural version of Ichabod Crane, long and narrow, with shutters hanging cockeyed from one hinge, and so shadowed by the woods behind that it was dark even at the height of the summer sun. The backyard was hardly a yard; it was damp dirt with a few scattered weeds and blades of grass. It was too dark there for anything to grow. Once, I set up a tent to air it out for Boy Scouts and after it had been staked, it smelled worse than when it had been rolled up. It was also covered in mildew. The Scout leader down the block yelled at me, calling me a liar and saying that I had not tried to dry the tent. In frustration, I quit the troop. Down the block too was the house with the concrete foundation where I fell playing touch football and jammed my knee so hard that I nearly ripped my kneecap off and needed a winding chain of stitches. It was a two-month-long recovery. Any more damage and I might have permanently ended my basketball career.

I didn't think too much about the larger world around me back then. I knew about things like Woodstock and Vietnam; I watched the moon shot in the summer of 1969. The year before, we saw film footage of the Detroit riots and the Chicago Democratic convention. I felt the bleak, collective sadness when Robert Kennedy died. But these were also things that happened a bit more on the periphery of my and my friends' lives. Growing up, I waited eagerly each Sunday night for *The Wonderful World of Disney* to appear on the television screen at 7 p.m., when Tinker Bell touched her wand to illuminate the castle of the Magic Kingdom with sparkly fairy dust.

That summer after fourth grade, like the summer before, I went to a religious camp down on Cape Cod. My grandparents and my mother drove me down and set me up in my bunk with six other children and a counselor. I loved the sand and the water and the clean Cape air and the quiet. Wakefield was the base of a triangle between the Route 1 corridor and Interstate 93, bounded by Route 128 and later Interstate 95, as this multitude of highways bisected the suburbs and converged, giant automotive arteries flowing to and from the beating heart of Boston. Wakefield itself was slow and quiet, with its large lake and Center Green for Fourth of July fireworks and an old-style bandstand. But the space around it was in constant motion. On the Cape, things held still, as in New Hampshire, except for the clockwork motion of the tides.

The camp was Christian, but it was the late 1960s, so many of the counselors had long hair and love beads dangling across their chests. They wore sandals and T-shirts emblazoned with peace signs. I was happy that we were busy. We made all kinds of objects out of Popsicle sticks and strung lanyards with gimp, the colored vinyl string. I was a master at the box stitch used to make eyeglass holders or key chains. One of the things we learned was sailing, and I took to the lessons, winning first place in the competition for small single sailboats.

At home now, we almost never attended church. I can count on two hands the number of times we went to services in Wakefield, or before that in Malden, or Revere. But at camp, we said grace before meals, our sun-bleached heads and peeling noses bowed over our plates. And the camp had religious speakers who came to address us in the amphitheater. I was either eight or nine when I heard one speaker talk about God. I've long since forgotten the words, but I sat, a rare thing for me, entranced. It was the first time that I had ever considered God, or given thought to the meaning of life. The first time that I began to consider, introspectively, the possibility of there being somebody, a higher power, who could help me. Walking to meals, steering my small boat out from the dock, before bed, several times a day, I began to pray, asking for guidance and support. I'd never ask for things, but for help. I'd ask for a clear mind, clear thoughts, good health, and the tools to do things better. When I was young, there would be times when I would pray, "Help me find a way to score fifty points or get the game-winning shot"; now all I usually ask is, "Please help my kids be safe" or "Please give me the strength to make good decisions."

Perhaps as a legacy of those summers, I've never felt that I needed a church. I go today, but over the years, I've always felt closest to God while riding my bike, pondering as the wheels move and the road passes beneath my feet the meaning of life and why I am here or how I am going to solve a particular problem. Whenever I have a conflict or need some guidance, a good bike ride or a run is how I handle it. I pray when I'm riding, when I'm running, when I'm swimming. I can dive into the lake or the ocean, feel the cold water surround me, hear the flap of bird wings, feel the blaze of the sun in a blue sky, and with each inhalation and exhalation feel connected to God.

I sometimes wonder if God found me in that camp because he knew what else would find me there.

I have purposefully erased his name from my mind, but I can remember how he looked, every inch of him: his long, sandy, light brown hair; his long, full mustache; the beads he wore; the tie-dyed T-shirts and the cutoff jeans, which gave him the look of a hippie. He had already finished college, so he was nearing his mid-twenties, and he was a counselor at the camp. He was somebody who was always kind of cool, walking with a real swagger. He was kind of funny, witty in his jokes and his put-downs—all that sort of crap. And he knew, from living around our cabin, that I was the kid whose parents hardly ever came to visit on parents' weekend, who got very few letters, who kept to himself, who could be alone even in a crowd. The counselors used to sleep with us, in their own bunks; we could hear their breathing a few feet away. Nothing happened while we were sleeping, although when we were changing into our pajamas or climbing into our sheets he might come in and stand too close, touching and brushing.

One afternoon, I didn't feel well. I went to the infirmary and he came, but the room was empty. I walked into the bathroom, and he followed. I was standing there with my pants down and he came right up next to me and asked me if I needed help, and then he reached out his hand. He began to play with me, his hands fondling me. I was ten years old. I took a step back and he had his own pants off, showing himself to me, saying, "This is what will happen when you get older," and he took my hand and placed it on him, to touch him. All I could think was: Oh God. Help. But he was my counselor, one of the few older men who had ever taken an interest in me. On the surface, he was a nice guy, he was fun, and I liked him. And he was my counselor, the person I was supposed to go to with any problem.

But going through my mind was: How do I get out of this? I was stuck, stuck with a predator, although I didn't know the word then. And I didn't know what to do. He was tall and he looked down at me and my hands and said, "Why don't you put it in your mouth?"

It was the Malden woods all over again, but I had no rock, nowhere to run. I saw the hair, smelled the thick scent of male sweat, and my stomach was close to heaving. My shoulder shook; my face must have looked alternately horrified and appalled. And I looked him in the eye and screamed, "No. Get away from me. No, no, no."

I yanked my hands away, and there on the institutional-tile floor, trapped by a sink, a toilet, and a mirror, I stood my ground. I tried to push my way by him, he stopped me, and I yelled "No," screaming at the top of my lungs. Up until that moment, I don't think he had imagined anything other than my giving in. Then suddenly, I think he got worried that someone else would hear me and enter the infirmary. He got nervous, glanced around, and abruptly stopped. Once he let me go, I raced outside. But there were several other times where he carefully maneuvered me into very similar situations, alone with him. It rapidly reached the point where I tried to avoid him. I was uncomfortable just at the sight of him striding into our cabin, even if every other one of my bunkmates was around. And he knew it. He got me alone one more time and told me that if I told anybody, ever, he'd hurt me badly, that he'd come to do it in the middle of the night, and that he would know if I said anything.

He leaned over, so that I could hear the clear threat in his tone. If you don't keep your mouth shut, he said, I will make sure that you never get the chance to say anything.

I did try to tell someone, my mother or my grandparents, but whatever I said, however I said it, I couldn't get the whole truth out or they didn't believe me. Finally,

I just said that I didn't want to go back to camp again. Ever. But I couldn't tell my mom or my grandparents why. When the next summer came, they packed my trunk and I was back again. He was still there, but I was in a different portion of the camp. I saw him once and yelled at him not to come near me. I stayed for the entire summer month, kept my distance from him, and nothing happened. The next year, I returned, and this time, he was gone. I was made an assistant counselor, and it was my best year. But I was always on my guard.

There were, I knew now, no safe havens, no one I could truly trust, just my legs beneath me, running, riding as far as they could carry me, and the slow motion of my lips, offering up a silent prayer.

Chapter Six

ALBION TO BROADWAY TO SALEM

The year I turned ten I started to grow my hair long. By the time I was eleven, it dipped down past my ears, came to rest against my neck, and then skimmed my shoulders until I could pull it back and stick it in an elastic band; my grandmother did just that, in fact, whenever I came to visit. The shorter front pieces flicked in my eyes, and when they grew too distracting I would take a pair of kitchen scissors and snip off the ends, until they fanned out in a slightly shorter, jagged fringe. My mother never told me to get a haircut, and so I was simply Scott, the long-haired kid, wearing my thick brown hair like a mane.

I had friends now in the neighborhood. Broadway Street, where we had the top-floor apartment, stretched between Albion and the train tracks running along North Avenue. Wrapping around beneath it like an asphalt dipper was Plymouth Road. My best friend, Jimmy Healy, lived there at number 54. It was only a little more than a block away, but the houses seemed as if they inhabited another world from the duplex on Broadway or the three rooms on Albion. Jimmy's house was big but not huge, but to me it looked enormous, ample enough for him and his brother, sister, mother, and father. There was a half basketball court in the backyard behind the fence and an inground pool.

We nicknamed Jimmy Healy "Hacka" that first summer, because every time that I would drive the ball to the basket, he would foul me—a hack—so I started calling him Hacka. We could play on his half-court for hours, shooting again and again until we stopped noticing time or the slow fade of the sun. The Healys always had new cars and a full fridge, and I so wanted to be a part of their family that one night I took a ring from my mother's jewelry box and went over to Jimmy's house to propose to his sister, Diane. I thought that if I married her, I could go to live with them. Jimmy's mother spied the ring, and I told her I found it in a box of Cracker Jack. She made a quick face and asked if she could look at it up close. When she turned it over in her hand, she realized that it wasn't a cheap faux-metal prize from a box of caramelized popcorn. It was a real ring. I think she called my mom and made me carry it carefully back home.

I could spend hours at Jimmy's, shooting hoop up into the net, or at the park or on my bike. I could spend hours anywhere but at 202 Broadway.

I whiled away hours at the pet shop on Albion Street, where I bought goldfish swimming in circles in plastic bags for under a dollar and watched cats, rabbits, hamsters, and guinea pigs play in the confines of their cages. And I began watching the school behind our new rented home. It was a long, low brick school, one story, with some frosted glass around the windows. The roof rose and fell in a series of peaks. Massachusetts winters were too much for most flat roofs; there had to be something to channel and melt the snow. The school sat up high on a hill, but I could easily glimpse the low field behind it through the mass of trees in our backyard and hear the carrying sounds of children running and playing.

It was summer, but the school was crowded—crowded enough, I realized, that it would be easy to join in, to play with the kids outside without being noticed. And so, late one morning, I hatched my plan. I slipped through the trees to the edge of the field and watched them, boys and girls my age, playing on the big field. A day or two later, I wandered out from the tree line just as they arrived on the grass, a big, long field with a baseball diamond at one end. There were games of kickball, baseball, and basketball. Like a circling wolf inching up on a new pack, I slowly tried to blend in and join their games. I stayed on the field as a bunch of classes rotated outside for recess and game time. The teachers watched me from the sidelines and the blacktop, wondering who I was. But their interest faded once the bell rang and it was time to head in. I never followed when the kids raced inside for lunch, but I wanted to. I was curious and always hungry. It was one of the periods when my mother was receiving some aid from the government. She was eligible for food assistance, and we got brown packages of government cheese, tubs of butter, and other food in cans. It was sporadic; she signed up in periods after marriages or between jobs, when she was working hard, doing her best, and trying to find some kind of stable employment that would support her and her two children—when she had to figure out how to be a sole provider again. Leeann was gone from the house in day care, in camp, in something, so until I went to that religious camp down on the Cape, the daytime hours were mine alone. At the school, there were games and food and activities to do in the classrooms, and I thought that I had discovered a great secret treasure, hidden just through the woods.

After a couple of days, when the regular students had gone in, I was standing in the parking lot, alone, dribbling a basketball. I was tall for my age, taller than all the other kids around, and my clothes were more than a bit too small and tight. I had on a sleeveless shirt and ratty shorts, dirty clothes that had been washed so many times the material had taken on a permanent state of dinge, and my long hair was falling in my eyes, looking as if it rarely saw a comb. A teacher walked up to me as I moved around the asphalt. I remember thinking that she was young and beautiful. I had seen her watching me. She smiled and began to talk, asking me questions, "Hi, who are you? Where do you live? What are you doing here?" I mumbled something and then she explained to me, "This is remedial summer school." I didn't really know what she meant, but after she told me, I was stunned. I had stumbled into a special session for kids who were struggling with their classes and lessons, who were forced to go to school, and here I was thinking it was the greatest place in the world. I looked up at her and I said, "Can I come here? Can I come to school here for the summer?"

I was, she later told me, the first and last kid she ever met who actually wanted to come to summer school, who didn't mind that everyone else was catching up.

Judy Patterson was a reading teacher, and she invited me into her classroom. She asked if I knew how to read, and then asked me to show her. She gave me a dictionary and told me to look up a word. I couldn't do it, so she showed me how to use the dictionary. Then she gave me another word and I found it. The next word was harder, something like "kinetics." I couldn't find that one, but she told me, "Well, it's a good try. You have the basic tools. Would you like to stay on with us? Just try it for a week? And come and have some fun?" I nodded my head, saying, "Yeah, that would be great." That afternoon, she spoke to the principal, who called my mom. The school didn't cost anything, and my mother was happy for me to have somewhere to go. I didn't need summer school—I was a B or C student—but I could play basketball on the blacktop and baseball in the field and I could run the bases in kickball. Inside the school, I loved the music classes and the singing, and Judy gave me extra work, advanced reading, which I also loved. And she kept watching me, more thoroughly than I had watched those kids from behind the tree line.

My summer school was my new elementary school as well. I studied in its classrooms, ate lunch under its pitched cafeteria roof, and ran on the slick, varnished floor of its gym. And I stayed at school long after the other kids had gone. At gym class and recess, I played baseball, usually as a pitcher because of my height and my arm. I could get the ball farther and faster than the smaller kids my age, and I had pretty good luck finding my target, although my throws tended to be wild, keeping the batters off-balance. But when I didn't have a team, I played baseball alone. On the weekends, or when school was out, I would walk over with my mitt and a ball and, just as I had done on Redfield Road, draw a chalk mitt on the brick wall and throw the ball against it for hours, catching it after it struck and bounced back or scooping it up like a grounder when it dropped and rolled along the cement. Or I might grab an old basketball that had been left behind on the court that bordered the grass field and play for hours on end. I dribbled and shot until they turned the lights out, and then I would walk home in the dark, moving through the woods toward the faint glow of lamps from the houses.

After a while I made friends, and other neighborhood kids would meet me on the field. Together we'd play baseball, basketball, and kickball, the rubber of our sneakers stinging against the rubber of the ball as it took flight through the air and the kicker ran the bases. Kickball can be played all year, and we played it even in winter, shedding our jackets as we ran around on the frozen fields. I wanted to be fast, so fast that when Jimmy or some of the other kids whipped the ball at me, it would sail past its target or if it found me, my feet would be solidly on base. We all wanted to be fast, but I wanted it most of all.

When the next summer came, I went back to summer school again. Miss Patterson was there too.

She watched me play basketball, and one afternoon, she said to me, "My boyfriend is the basketball coach for the eighth-grade team. I'd love for you to meet him sometime." I said sure. Brad Simpson was a tall, handsome guy. He watched me dribble and shoot not once, but again and again in the afternoons. When I paused, my too-small shirt damp with sweat, he would say that he couldn't wait to get me up in the eighth grade. But, he added, "I don't know if you're good enough, Scott. I don't know if you can handle it." He'd shake his head slightly and tell me that I needed more skills, and then he'd say, "I want you to do this," and he'd show me a shot or a fake or a drill, or "I want you to do that." And he'd show me ways to dribble. I know now exactly what he was doing, but back then I took it like a challenge, a doubt that I would measure up. I felt I had something to prove. One day, Brad gave me a ball of my own, a regulation-size basketball, and I dribbled it everywhere.

I dribbled the ball during lunch. If I was walking downtown, I would dribble the ball the whole way. Wherever I went, so did the basketball, vibrating up and down beneath the palm of my hand. I could switch between my left and right hands; I could dribble it around my entire body, passing it from palm to palm in one seamless motion. I dribbled the ball so much that the pebble skin wore away until it was smooth and Brad gave me another one.

When I couldn't bear the sounds around me, I'd drown out the noise with my basketball.

Most nights, I would take my basketball to bed with me. I would lie in the dark, sometimes crying, sometimes thinking, but most of the time just talking to my basketball, and I would fall asleep with it in the crook of my arm. It was nothing more than a nylon carcass and a butyl rubber bladder pumped full of air, but it seemed to be nodding or occasionally whispering back to me, a sage sphere listening in the darkness, absorbing my secrets and my despair. I locked myself away under my covers, thinking, "What's the point? Is this it?" and pleading to the silence, "There has got to be more."

I was a tight-lipped kid, and I told no one anything, not the truth about my father, not about the swallowed-up emptiness I felt, not about my mother, whom I battled like a boxer in the ring, not about the stray images that crept in like leaden New England clouds, the tile in the camp bathroom in Cape Cod, the dense Malden woods and the rock in my hand. From daylight to dusk, I ran the bases, broke away on the court; I filled my days with motion because when I was moving, when I was diving for the ball, cocking my wrist to shoot, swinging a heavy wooden bat, there was no time to think. Like a reflex deep inside, my muscles took over, my mind focused only on guiding them, on the elements of breathing, on leaning, shifting, and dodging. But at night, under my covers, there was only quiet solitude. I couldn't escape my thoughts; I couldn't escape anything. I was trapped on the second floor.

I don't remember when my mother and I started fighting. In the beginning, it was over whether I had cleaned my room or where I was, or how I had failed to come home on time. The starting point was irrelevant; it was the end that mattered. She would yell, and I would yell back. The slightest disagreement could easily escalate into a full pitched battle, with our faces inches away from each other and then with her smacking me, with a towel, a belt, or the dreaded two-by-four. Finally, when I got physically big enough, I shouted, "Don't ever hit me again because I am sick and frickin' tired of it." After that, it became a battle of words, although she might give me a shove or a push. It all blends in like so much loud noise rattling around in my ears.

She was tired, my mother. She was working at dead-end jobs and coming home to an apartment that she had to clean, to laundry in piles, bedsheets, towels, dirt and baseball grit tracked across the floor. Even though we lived upstairs, not all of the dirt came off before we made it in. She had rent payments, car payments, car shop payments because used cars were all she could afford and they were always troublesome. She had grocery bills and she had to buy me new clothes because I couldn't stay small long enough to fit into anything. Her friends were all married, and they ran errands and did the food shopping when their husbands were at the office. My mother had to do it after work or on the weekends. If something broke, she had to fix it. There was no backup, no other shoulder. She never made plans, never talked about the future—hers, Leeann's, or mine. Even something as simple as going to the beach for an afternoon was a spur-of-the-moment thing on a Saturday morning. She never looked back or looked forward. She was a get-through-one-hour, get-through-the-next-hour type of person. She worked hard for us, and any little bit of money that she spent on herself went to pay for exactly two things: her cigarettes—Marlboro 100s that she bought by the carton—and alcohol.

Sometimes I hid her cigarettes and she would be furious, tearing the house apart looking for them. And sometimes I went after her booze. A few times, I emptied the bottles down the sink, hearing the steady glug, glug as the liquid was swallowed up by the drain. More than once, I diluted the contents of a bottle with water. She liked vodka best—her drink of choice was a vodka tonic. She bought it in economy gallon jugs with Russian-sounding names like Popov, and when she had worked her way through the first third of the alcohol, I could add the water without her knowing. Just a cupful or two at a time into the clear liquid of the big gallon jug, and she could only wonder why her drink seemed to lack its usual kick. Of course, probably all she did was pour more.

Most things were worse when she drank, especially the fighting.

She didn't drink all the time, but she was constantly around alcohol in her waitress and banquet-serving jobs. When she went out with her old high school girlfriends, it was invariably to a bar or a cocktail lounge. And everyone drank back then. The men who rode the train to and from Boston each weekday came home and poured themselves two fingers of scotch or kept jars of pearl onions or olives to drop into their gin and vermouth. A few years later, when I got a job in a liquor store down the road, I spent my late afternoons and Saturdays wearing a path from the stockroom to the parking lot, carting out cases of beer. Teenage kids drank beer in the cemetery; it was hardly rare to see half-crushed cans of Pabst Blue Ribbon or Budweiser glinting next to sedate carved headstones.

My mother kept her bottles under the sink or in the cabinets. Sometimes she would hide her alcohol, but I could almost always find it. My nose grew so attuned that I could smell it before I even walked into the room. And I didn't have to hear the soft click of the cabinet or see the empty glass in the sink to know when she had been drinking. From almost the first sip, her entire personality changed. The moment that initial jolt of alcohol was absorbed into her veins, her transformation would begin. When she drank, everything about her sharpened, especially the knifelike barbs of her tongue; she was someone I did not want to be around.

She did not often get sloppy, falling-down drunk.

But at times I did come home to find her passed out on the couch, her body splayed like a fighter on the losing end. I've seen her stumbling and vomiting—into the toilet or the sink, whichever was closer—and smelled the sickly stench long after it had been washed down the drain. It was probably why when she went out to bars with her girlfriends she left her car at home.

My mother and I battled the way she battled with her husbands, especially with my father. We shared that undercurrent of tension, smoldering below the surface, which at any minute could be kicked into high flame. She didn't like my choice of friends, especially the older kids. She didn't like the girls who had started to hang around, especially the sixteen-year-old who would show up under my window at 1 a.m. and call my name, asking me to come out with her. She didn't like how I disappeared or was late or didn't show up when I was supposed to. Mostly, she probably didn't like that somewhere between age ten and age twelve, I had slipped beyond her control. I was bigger than she was now; I towered over her, and my voice was rumbling. I could and did pass for a high school junior or senior, not a preteen, and that made it easier for me to take off on my own. Sometimes, I would just up and disappear, and it was like throwing in her face all those other vanishing men. So we fought our pitched battles, screaming, prepared to come to blows, and we did it right in front of Leeann.

Sometime after I entered junior high, we moved again. Off Broadway Street and the perpetual mold of its too many trees, away from the peeling paint and shutters hanging lopsided by one hinge. Away from my third-floor bedroom where one night, sweaty and quick, I lost my virginity to an eighteen-year-old girl from up the street whom my mother periodically hired to babysit for Leeann. I had real girlfriends later on, but after that night, it was years before I had a serious one. We moved to another top floor, the second floor of a house that had been chopped into quarters: two apartments on the top, two on the bottom, my room under the eaves. The street was named Salem, like the town of the witch trials and the courthouse with Judge Zoll. It was, ironically, only a few blocks down from Eastern Avenue, where my grandparents lived, but it might as well have been a world away. When Gramps retired at sixty-three or sixty-five, he and my grandmother retreated back to Portsmouth, New Hampshire, full-time. They bought a nice Federal-style house that had probably once belonged to a ship's captain. Gramps could oversee his rentals, be near the water, and inhabit the flinty world where he was most at home. There was no train whistle to remind him of his clock-managed life commuting into and out of Boston. But their absence meant that I couldn't ride over to see them. I couldn't run away, not even to the red vinyl chair in Gram's kitchen. We still drove up to Portsmouth—it was only about an hour's ride—but it had to be a planned excursion. The spontaneity was gone.

The Salem Street apartment was smaller. But there was a basketball hoop in the driveway, and I hammered a nail and put another hoop up on the utility pole along the dead-end road across the way. I moved my mother's car to shoot, when I was not driving it on small excursions. One night she went out, leaving me with Leeann. I had a friend over and we decided to go get some beers at the liquor store with our girlfriends. We hopped into the Impala and left Leeann in the living room with the television, watching Sonny and Cher. I came back with the car and parked it perfectly. Or so I thought. It was actually too perfect, too close to the curb. Mom recognized that the car wasn't where she left it and busted me for getting into it. I told her that I was just moving it so that we could play hoop in the driveway, which might have been a good enough lie, but the car was too far away and too perfectly placed and she didn't believe me. She was mad, furious. "Do you know what you've done? You took the car and left Leeann." I tried to act contrite, to show that I knew when I'd crossed the line. The thing was, she never found the beer or any sign of the girls. As with most parts of my life, she didn't know the half of those other things.

I wanted to go out now, almost every weekend. At the local YMCA and the town hoop courts, I had met Bobby and Jay Moore, two basketball-playing brothers. Bobby towered at six foot five and was a budding high school star, and Jay, six foot one, was coming up fast behind him. Bobby and Jay played every sport, including street hockey and ice hockey when the lake froze and boys rushed down and strapped on their skates to glide over the pebbly, bumpy ice. And they played basketball. Bobby's nickname was "Ace," and I became "Deuce," because I could hold my own on the court with him, and I sought both brothers out, all the time. I wanted to be just like Bobby; I think he kind of tolerated me. There was always a game at the Moores' house, a fierce contest waged in their tiny driveway. Bobby would play from early morning until past dark, and when they put out lights on the side, he kept shooting long after the sun went down.

Wakefield in 1971 was a place where the parks were crowded with kids, where there was always a pickup game or someone with a mitt and a bat. Mrs. Moore turned her boys out in the morning and told them not to come back until supper, which she served when the sun was almost down. "Out of the house," she said, and the rule was ironclad. If you came to play at the Moores' and you were thirsty, you turned on the outdoor faucet at the side and sucked down the stream of clear, fast-flowing water. Kids came and went from that driveway at all hours, and when one crop grew exhausted, another appeared. There were games of one-on-one, two-on-two, and three-on-three, or Bobby might just play one-on-two against two younger kids. There were no age limits, and the rules were clear: winners stayed on the court until they lost; losers sat down, and only then could new players rotate in. The space itself was so small it wouldn't even be a regulation quarter-court, but the narrowness just made the game more competitive and physical. Rather than spread out for a pass, we converged on the net and on the ball. It was a game of elbows, fakes, side steps, and dunking.

Balls landed in the net or bounced against the neighbor's house, which pressed up against the edge of the driveway, barely big enough for a full-size Oldsmobile to park in. I could hear the games from down the block as I ran to go play, eager to take Bobby on. I heard the slap of sneakered feet on the blacktop, the rhythmic pop of the balls, and the cries to pass, the shouts at a score, or the wail of despair when a ball rolled around the rim and then tilted sideways back to earth, failing to drop in. I started showing up at the driveway on most days, sometimes waking Bobby up and badgering him to get out of bed so that we could go shoot. But no one turned me away.

I was almost as tall as Jay, and although the brothers were nearly three and five years older than me—a veritable lifetime in the world of teenage boys—I could keep up with them, not just breaking away on the court, but faking, passing, rising up on my toes to shoot, and letting the ball sail through the air, released at the perfect moment, with the perfect wrist extension. Bobby was always teaching me to play better, always challenging me to be better. I was usually the first one picked for teams when we played. One by one, other players tired and dropped out, but not me. I was determined to outrun all of them. I earned their respect on the court, and they found it funny to have me tagging along after them. They were my surrogate big brothers, my role models and mentors, a point to set my compass by as I tried to figure out how to stop being a boy and start becoming a man. But to them, almost forty years ago, I was just the scrappy kid who was always hanging around and who never sat still.

Later, we played together in summer leagues, traveling to Andover and Newburyport and Boston, lacing up our sneakers and trying to whip other teams. But mostly, in the beginning, I was a fixture at their house. I was embarrassed to have people, especially older kids, come to mine. It was small and spartan, and there often seemed to be some type of conflict. I remember a number of times when Mrs. Moore sent me home with a bag of clothes that she said Bobby had outgrown. I ended up wearing long coats that hung down past my knees because that was the style Jay wore.

On weekend nights when their parents had gone up to Maine, Bobby and Jay had parties: they grilled hamburgers and hot dogs, played hoop, and drank beer. The drinking age was still eighteen, and someone always brought beer. In addition to the games in the driveway, some of the older kids would hang out on the tiny porch of their brown-shingle home. As a great joke, Jay and Ace invariably introduced me as a sophomore friend of theirs who went to Melrose High, over in the next town. I hung out with their friends, and occasionally drove them home when they had too much to drink, even though I was only thirteen. And I tried to kiss the high school girls, who thought I was cute until their younger sisters told them how they sat across from me in the same classroom in junior high. I took sips from a beer can and worked at acting cool, until it got close to my curfew time, 10 p.m. At five to ten, I would take off running, a fast sprint up and down the rolling hills of Wakefield, to be home on time, or two or three minutes late. I trained myself to run a five-minute mile to make my curfew on those evenings.

Hanging out with a pack of older kids was also partly how I ended up in a car at the Liberty Tree Mall on that hot July afternoon, and from there, in Judge Samuel Zoll's courtroom.

Chapter Seven

JUDY, BRAD, AND THE JUDGE

About the time we moved to Salem Street, my dad was getting divorced again. He moved out of his big house in downtown Newburyport and into an apartment nearby, and now Robyn and Bruce were a bit more like me, vying for his time on alternate weekends. The difference was that he saw them more consistently. I was still, for the most part, a child in absentia. But that made it possible for me to imagine a very different life with him. I had taken a few brief tastes, and I thought that I wanted more.

I was aware that in Newburyport, my father was a man of some consequence. He was an elected member of the city council and I think he served as president of the council at one point. He was also one of a few residents who was trying to rescue this once prosperous town. In its glory days, Newburyport had been a small but wealthy port. Elegant captains' houses rose on the hill aptly named High Street, and African and Native American slaves labored behind older mansion doors until after the end of the American Revolution. Closer down toward the docks the houses crowded on top of each other, but they still retained a kind of faded elegance. Newburyport was, in its prime, a place of shipbuilders, merchants, and traders, wedged on a thin strip above the marshlands at the tip of the Massachusetts coastline. New Hampshire was a ten-minute ride away. On the town docks, built out into a channel of the Merrimack River, barrels of thick West Indian molasses were unloaded and then turned into heady rum in the distilleries that lined the Market Square. Its small band of citizens held the first Tea Party rebellion in protest against the British tea tax, and it was later a hotbed of abolitionism and a vital link in the Underground Railroad.

But by 1970 the historic downtown, largely abandoned for the ease of strip malls and bypassed by the snaking highways, was scheduled to be razed. Some people even wanted to put another strip mall along the water. The town had fallen into such a state of disrepair that owners could hardly give their properties away. Only at the last minute did the wrecking ball stop, and the downtown was saved. Instead, a federal grant helped pay for renovation and restoration, and my father had helped to secure that money. I knew that my father was involved in the restoration efforts, but what I knew most was that he had joined with a partner to buy a boat.

Christened the *Sabino*, it was a 1908 coastal steamer that had plied the Damariscotta River of Maine and traveled among the scattered islands of Casco Bay. It had a coal-fired engine and could hold fifty-six passengers. One summer while I was in junior high, my father let me spend a week with him working on the boat. I scrubbed the rusty toilet bowls, collected tickets, fed the coal for the engine, and hung off the lower reaches of the decks as the steamer chugged along the river channel, the spray from the churning waves coating my dangling legs and arms. I loved the feel of the boat and the freedom of being away. At night, I slept on a living room couch amid the photos of Robyn and Bruce.

Back in Wakefield, I imagined Newburyport. And when my mother screamed in anger and frustration, "Why don't you go live with your father? See how you like it!" it didn't help that I had already begun running away.

In truth, I had been running since my days at Aunt Nancy's on Redfield Road, when I would hop onto my bike and pedal to my grandparents' house. Before that, I had run out the door on Eastern Avenue. I was a runner, a bolter; when trouble hit, my first instinct was to move. When my mother and I battled, all I wanted to do was run. And I did.

At first, I'd run to friends, and I'd ask if I could stay the night. Sometimes, I took clothes; other times, I arrived with whatever I had on. The parents would call and say that Scott is here, and there would be a cooling-off period before I went back home. Then came a particularly vicious fight when my mother threatened to break the few things I owned, including what I treasured most—my trophies. In that moment, all of Wakefield seemed too small. I called "Sorry" to Leeann, who was still in the apartment, looking stunned and trying to sink into the woodwork as our small war was waged around her. This time, I was thinking, I would be gone for good.

My feet thundered down the stairs and out the door.

With only the clothes on my back and $1 in my pocket, I raced for my bike, and I rode. I expected my mother to chase me. I expected that any minute I would glance back and see the white Impala sailing over a hill and glimpse my mother, her hands clamped to the steering wheel and her foot gunning the engine. I pumped my legs furiously and fixed my mind on a new destination. I would ride to Newburyport, to my father's home. It was thirty-five miles away. I followed Salem Street as it snaked out toward Saugus and Route 1, the highway mecca of the Hilltop Steak House Restaurant and Kowloon Chinese, Hilltop's giant cactus facing off against Kowloon's fearsome imperial dragon. Farther down along the commercial strip, the once-swank Caruso's Diplomat, where my mother had worked, was slowly being overcome by strip stores and discount-tire marts.

I merged onto the busy highway, riding like the wind. I thought my mother would follow me. My only choice was to be faster, to outride and outrun her. The cars and trucks whipped by. I passed Newberry Plaza, the Golden Banana strip club, with its windowless walls and neon sign, a quick-lube shop, and a mobile home park. Some ten miles out the traffic fell away, merging north onto the multilane interstate. Route 1 broke off by itself and became quiet, hilly, and green. It had always been the Newburyport Turnpike, but this stretch felt more like an old colonial road, with the roar of the trucks largely gone.

There were apple orchards and the spreading grounds of the Topsfield Fair, the oldest agricultural fair in the nation. I had been there as a child with my grandparents, to see the cattle, sheep, goats, and horses paraded in the show ring. But all I could think of now were the hills. The road into Topsfield rose and dipped like mountainous camel humps. I felt the burn in my legs as I pushed up each hill, and then the brace of wind as I coasted down. Once, on the narrow road, a truck passed with a chain wildly swinging off its back. The metal whipped with the wind and nearly clipped me on the side of my bare head. But I kept pedaling.

Finally, on a stretch of open road past the thick woods, where it was easy to imagine woodland Indians silently stalking their prey, I stopped at the Agawam Diner, a silver building planted by the roadside, low to the ground, with bright awnings. I had my dollar in my pocket, enough for two Hershey bars, and I was starving. I paid the cashier with my matted bill, tore open the wrappers, and shoved the chocolate into my mouth, barely chewing, barely tasting anything. I was so desperate for the sugar, I didn't think about drinking anything.

I pedaled past the marshlands on the south of the

town, past the turnoff for Plum Island and onto High Street, then down to Prospect, which was bisected by Fair Street, then Federal, and then Lime. It was dusk when I arrived; the ride had taken me all afternoon. Panting, exhausted, and parched, I stepped down from the bike and knocked on my father's door. Nothing. The place was black. It had never occurred to me that he wouldn't be home. I got back on my bike and rode to the wharf, every muscle in a state of rebellion, to where the *Sabino* was tied up. But the steamer was empty, its coal engine cold. My hair was plastered to my head with sweat; my face was red and windburned. Night was falling, and it was starting to rain. I was near tears. One of my father's buddies spotted me, and he must have called a bar or somewhere because about fifteen minutes later, C. Bruce Brown appeared to take me to his home.

He called my mother to say, "Scott's here," and she was utterly stunned that I had reached Newburyport in an afternoon. She had just let me go, assuming I'd turn up somewhere in Wakefield, never actually contemplating where else I might have gone. My dad fed me, told me to take a shower, and gave me a pillow and a blanket for the couch. It was all he had to offer, and I don't know what more I was expecting. My legs were like rubber from the thirty-five-mile ride, a stew of lactic acid, flesh, and bone. I had found a way to burn off some of the anger; for me, the path to inner peace was through sheer physical exhaustion. By shutting down my body, I could also shut down my mind.

I stayed at my dad's for a day or two, and then he strapped my bike to his car and drove me home. I was never going to live with him. I was never going to be more than a week in the summer, a day trip, or a quick overnight on the weekend. He had already made that abundantly clear on the *Sabino*, when I, the toilet scrubber, deck cleaner, and eager apprentice coal man, was introduced by my father as "Scott, my son, who lives with his mom."

I ran away to Newburyport one more time after that, making the same perilous ride along the highway and then the older, hillier Route 1. I rode again like a person possessed, convinced that my mother and the Impala were behind me the entire time. But except for a few cars bound for other destinations, I was alone.

Even back in Wakefield, I did not stop running, although after that last trek to Newburyport, my excursions stayed closer to home. My mother and I kept fighting, and now I would pack up my few things in a footlocker, hoist it onto my shoulder, and start walking, a mile or two at a time, down Salem and over to 22

Valley Street, where Audrey, one of my old babysitters, lived. She was younger than my mom, she had her own kids, and she was a free spirit who wore 1960s-style beads that clicked lightly around her neck. Her silver-streaked hair hung down to her waist. It swayed and moved of its own accord whenever she turned around. I'd stay at her small place for a few days or even a week until the furor had subsided, and then she'd drive me and my trunk back home. Sometimes I'd go to the house of one of my mom's childhood friends, Judy Vining. She had two kids, Lenny and Dana, and we had a lot in common, and they always welcomed me when things got tough and I needed to run.

At home, I papered my walls with posters of my basketball heroes and KISS and stared up at Gene Simmons's caked makeup and oversize tongue. I had a mother who found the world more tolerable with a generous pour of Popov and a lit Marlboro, but who often couldn't find me tolerable, and who hated the disappearing man who had given me my last name.

I lost myself in school. At the start of junior high, I joined the seventh-grade basketball team and made cocaptain. The next year I was on the eighth-grade team, again as cocaptain. I was responsible for leading the calisthenics and the drills, and for making sure that everyone showed up for practice. At the start of the game,

I was the one who had to go up to the referee to get instructions. I had to make sure that the balls were on the court. If practice was canceled, added, or changed, I was at the head of the call list. It was my responsibility to make sure that everyone else on the team knew. My coaches were teaching me responsibility and giving me structure, although I didn't recognize it then.

On the court, I was not always the fastest kid, but I was considered one of the hardest workers and one of the toughest. I never quit. I was always dribbling and driving to the basket, working on my game. I wasn't much of a passer because when I got the ball, all I wanted to do was move, head down the court to the basket, and rise up off my toes. I played and I studied. Sometimes Brad Simpson, the eighth-grade coach, would watch me, even though I was only on the seventh-grade team. I was a lefty, constantly dribbling with my left hand. One time, he told me that if I wanted to get really good, I needed to be able to dribble right-handed. After that, everywhere I went with the ball, even through the village green downtown, I dribbled with my right hand, willing the muscles on that side to become as quick and as strong.

That summer after seventh grade, tragedy struck our team. Our seventh-grade coach and his pregnant wife were killed in a fiery car crash. Their funeral was on a weekend. It happened to be one week that I was with my dad, working on the *Sabino.* I told him about the wake and the funeral, and how everyone on the team was going; it was particularly expected of me, as a captain. But my father wasn't interested in driving me back to Wakefield. It didn't fit with his plans. So I didn't go. I heard from Brad and Judy and from the assistant coach how disappointed they were in me. I felt like crap. I hung my head and said nothing. I never told them why: that it was my dad; that it was his decision, not mine; that he would not let me go. I didn't like excuses and I wasn't going to make them. I should have found my own way home. Even now, I have in my scrapbook the yellowed news clippings about the accident and the funeral, which my grandmother pasted in, a page or two down from my commendation certificate that Coach George Geyer had signed.

In eighth grade, Brad Simpson was my coach and Judy Patterson was my social studies teacher. She never seemed to mind how many times I raised my hand. The fall of the year coincided with the 1972 presidential election between Nixon and McGovern, and she organized a mock vote in our classroom. I wanted to participate in everything, from writing the ballots to standing at the ballot box when the votes were cast to counting up the tallies.

But Judy also was not afraid to rein me in. Once, when she overheard me walking down the hall and making a cutting comment to an unpopular girl, she grabbed me by my long hair, pulled me into her classroom, and slammed the door shut. "You know what?" she said. "You've got everything in the world going for you. You're tall, you're good-looking, you're athletic. You could be smart if you put your mind to it. But you're a jerk. How dare you say that to that poor girl? How do you think she's going to feel now for the rest of her life?"

The question hung there. It was as if she had just smacked me upside my head. I was the kid who always felt like a loser, who felt I had nothing going for me, and Judy said I had everything in the world going for me. She was holding me accountable. I looked out at her from under my unkempt hair. I veered from wanting to be a rock star to wanting to be a starter for the Celtics, although I had never been to a game where the crowd sat on anything other than pullout bleachers or where the floor varnish was not being slowly stripped away by PE classes and full-court basketball. I listened and I felt ashamed.

And I knew that she was the first person in a long time who actually cared. In one of the universe's strange ironies, she and my mom had been given the same first name.

When classes ended, Brad took over on the court. He held practice every day after school and sometimes on Saturdays in the morning. Back then, eighth-grade basketball was a pretty big deal. Reporters came to see us play and wrote weekly dispatches highlighting our games. There were local newspaper photos and boxes of stats showing off the high scorers.

Brad never talked about what was in the paper; he simply directed every last bit of his towering intensity into making us play to win. He was tall, with Irish good looks, and a face that turned plum red when he was racing around the court. Most days after practice, I would play him, one-on-one or two-on-two with his assistant coach, Jay Connolly, and other kids from our team, such as Bob Najarian, Billy Cole, Mark Gonnella, or Hacka Healy. Brad was bigger and stronger, if not always quicker, but for years, he won every time, and I watched all his moves, studying him. Jay was much taller, a bruiser, using his elbows, pushing to the floor; no-nonsense play was his style. He made us all tougher. When we came close to beating Coach Simpson and him, they stopped playing us. After we got better, we would have to beg, tease, criticize, and humiliate them to goad them into playing. Coach especially, after all those years, wouldn't give us the satisfaction of winning.

He, in turn, expected everything from me—rebounding, passing, scoring. He drilled us all, every day, and he took no crap from anyone. Instead, he goaded me with possibilities. "Hey, next year you're going to play as a freshman," he would say. "Who knows, maybe you'll be able to play varsity." Except that no one played varsity as a freshman. Schools did not allow freshmen to play varsity back then, and most sophomore players sat on the varsity bench. It wasn't done. But he planted the seed. I wanted to play varsity as a freshman. I wanted to be the first one. I could already feel the polyester jersey on my back. I did his calisthenics, ran his wind sprints, and did his drills, and then we'd play, the entire team, two-on-two, three-on-three, four-on-four; then we'd scrimmage. Then he'd line us up for shots from the foul line. If you missed, you ran five or ten laps around the gym. If you made it, you were done. It was precise and disciplined; for two hours, there was no chaos. It was only the blast of the whistle and the voice of Coach Simpson.

One day he told me, "You'd be a heck of a lot better player if you could actually see the ball. You look like a girl with all that hair." My hair was in my eyes, down my neck, and wild. When it got unbearable, I'd cut a few strands with the kitchen scissors. Before practice, I ducked into Judy's classroom and told her that I had to do something about my hair, so that I could see the ball. "Please," I said, "cut my hair." Judy was more than a little surprised and asked about my mother. "My mother won't even notice," I told her. "Please." So she took the scissors out of her desk drawer and took some off the front and the sides, the back and the top, until it was possible to see my face again. After the next game, I had her take a little more off, and I brought another kid from the team for a trim. After the next game, I brought in two more kids for a Miss Patterson trim.

We played a lot of tough teams: Burlington, Woburn, Lexington. We had eighth-grade groupies, girls who came to watch, and our parents. My mom came to nearly all my games, most of the time with Leeann, who, to amuse herself, tried to turn cartwheels alongside the gym wall. My mother sat there, all frosted blond hair and frosted lipstick, and I don't remember if she kept her coat on. But I knew she wasn't quite like all the other moms. One thing my mom always took with her was the shouting. Just as she'd shout at me or at my dad, she'd shout at the referee, questioning the calls, calling him blind or dumb. I could hear her, I was always very glad that she was there, but at a certain point, I just stopped listening. I was proud of my mom and her support, but for an eighth-grade boy, it was also at times uncomfortable and even embarrassing. So I tuned everything out. It was only me and the ball. Just as I could train myself to dribble with my right hand, I could train my ears to block out every other sound.

I distinctly remember that my father was supposed to come to one of our early games. He promised he'd be at our third game. I told Coach Simpson to expect him, that I'd like them to meet each other. But my father never showed up. I spent the first half scanning the bleachers, but no sign of him. The second half, all I did was play. Now he was supposed to come to the sixth game. Again, I scanned the stands; again, nothing. And two games after that, and then the next game. By now, Coach Simpson knew not to say anything. The other boys had their fathers. I had my mother in her frosted makeup and my grandmother, who in New Hampshire cut out my newspaper clippings. Coach pushed me harder. More drills. More rebounds.

Finally, for the second-to-last game of the season, my father came. He had his latest girlfriend on his arm. My mother was also in the stands, glaring at him when she wasn't glaring at the referee. Afterward, I took my dad over to meet my coach, but my dad was with his girlfriend, and my mother was standing right there, seething. The air was electric with the tension. Coach Simpson felt it. Everyone was stilted; everyone's mouth was frozen. I was the high scorer, the team captain, but I had no answer to my own family battles on the sidelines. My father said something about how he used to play in his glory days and my mother made a crack in return. My father and his girlfriend made a quick exit. I, as always, would be going home with my mom, and listening to her string of comments about my dad.

I was angry, angry all the time. I was angry at everybody, but especially at my mom and dad and at the games they played with me and my emotions. I was angry that I couldn't even count on them to hold it together for a simple basketball game. It was embarrassing always to have to make excuses, excuses as to why my parents weren't together, why my dad wasn't coming, why I didn't have nicer clothes or any extra cash. I had to make excuses or apologize when my dad or mom didn't show up at a school event or function, at something for fathers or for families. I had them all in my head: my dad got caught at work, or I forgot to tell him, or he's away, or he's moving. Or sometimes when he wouldn't show up, I said nothing. I refused to be a whiner. But while I could run to Bobby and Jay's house, or to Audrey's, or Judy Vining's, or get a job at Dunkin' Donuts making doughnuts or cleaning the toilets and the grease traps in the fryer, or work at a liquor store, I couldn't see a way out. I was the kid looking for a fight, and if I wasn't looking, I certainly wouldn't shrink from one. And I was going on to high school. I was leaving the cocoon of Brad and Judy on the court and in the classroom.

That was, of course, the summer when I ended up in a car with a trunkful of boosted records that I had stuffed in my overalls, earning me a summons to Judge Samuel Zoll's courtroom.

I could have ended up in so many other spots. If I had been caught at the Square One Mall in Saugus, I would have been sent to the Middlesex County Courthouse. It just so happened that Liberty Tree was over the line in Essex County, so I was sent to Salem. If the timing were different, I could have ended up with someone other than Judge Zoll. He had been appointed to the bench only that year. Prior to that, he had been the mayor of Salem, a state representative, a city councilman, and a high school teacher. A Korean War veteran, he had worked his way through law school at nights to get a degree. If I was going to get busted, it was, in retrospect, perfect timing.

Judge Zoll's chambers were magnificent, a reminder of the grand riches of seafaring Salem. But his six-foot-four frame almost seemed to dwarf the chambers. I stood when he stood, and I felt the size of him. I looked at him, but I cowered too, keeping my eyes down. It was as if at that moment he was deciding far more than the adjudication of my Salem District Court summons. He motioned to me with his oversize hands, and I sat in my stolen suit, and my shirt and tie. Thinking back now, I'm amazed my mother never asked where the suit came from or how it magically appeared on me that morning.

I tried to look the judge in the eye, to be respectful, to use all the manners that my grandmother had taught me, as he studied my file. I had no way of knowing that Judge Zoll had a houseful of kids at home, that he knew the names of every artist and album that I had stolen. And Judge Zoll was a basketball junkie.

He began by asking me questions. He asked me about music, about sports, about my studies. "I know you live in Wakefield," he said. "Tell me about yourself. You obviously like music. What kind of music do you like?" Each time I answered, I tried to look him in the eye, as Gram and Gramps had taught me back when I still listened, and as my coaches demanded I do whenever they were speaking. I tried to be respectful.

He began asking me more about sports, and I began to relax. These were questions I could handle, questions I loved. I could talk about sports for hours in the cool of his chambers. Then the judge asked me if I was a good basketball player. I looked straight at him and said, "Judge, I'm an excellent basketball player." Almost forty years later it sounds cocky, but in that moment, I sounded confident, and I caught the judge's attention. He said, "Wow. That's fantastic. How many points do you average a game?" "Twenty to thirty," I told him.

Years later, when I spoke to him about our first meeting, he recalled that he had seen kids come through who were disrespectful and slovenly, who mumbled their answers and were ready to game the system. But I didn't even know there was a system.

He asked me about my studies, "Scott, are you a good student?" I told him I was, adding, "I'm really good in some areas, and in other areas I need a lot of work. And I always try to do better." Then he started asking about my family: Did I have brothers or sisters? And I told him about Leeann and Robyn and Bruce. It was too complicated to explain that Leeann was a half sister, since she had the same name and lived with me, but I told him that Robyn and Bruce were half siblings and that I didn't see them too much. And he asked more questions, and he did something few adults ever seemed to do. He listened. Aside from my coach and my grandparents and Judy Patterson, he was the first person who seemed genuinely interested in who I was and what I was doing. Had he been distracted, rushed, or perfunctory, there is no telling what would have happened. Had it been a decade later, he probably could never have spoken to me in his chambers, alone. But on that particular morning, he looked down at me with his massive frame in that big leather chair and threw me a lifeline. He wound his way back to basketball and asked me if my sister and my half brother and sister came to see me play. "Yeah," I answered. "A lot of times they try to."

"Do they look up to you?" he asked. "Yeah, they look up to me. I'm the guy who tries to keep everybody together." And he said, "Wow. That's great. How do you think they will like seeing you play basketball at the local house of correction? Because that's where you're going. You're on your way to jail right now as evidenced by the way you went in and stole these records. You really didn't care about the businesses that had to work hard to pay their employees and the fact that you took something that wasn't yours." It was as simple as that.

I know now how I seemed to Judge Zoll on that morning: lost, poised to go horribly wrong, but with potential. And in those moments he decided that I was worthy of help. My sentence, as he handed it down, was "to write a 1,500 word essay on 'How I disappointed my brother and sisters and how I think they would like to see me play basketball in jail.' "

He turned to the probation officer—Mr. Burke, I think his name was—and added a warning. "Scott, Mr. Burke is going to be my English teacher here. He's going to check that essay for grammar, sentence structure, and punctuation, to make sure that it's a thoughtful piece of work. And if it's not good, you're going to be in serious trouble." I looked him straight in the eye and promised, "Judge, don't worry. It'll be good. I'll make sure it's good. I won't let you down."

I slaved over that essay with the same determination that I pounded my baseball into the concrete wall or my basketball against the backboard. I sat in my room on Salem Street and wrote and recopied, wrote and recopied; it was hard for me to write cursive with a pen because I'm left-handed. It gave me time too to think about my coaches, about Brad and Judy, and how I had disappointed them. I called the probation officer a couple of times with questions, asking if it should be pen or pencil—should it be double-spaced, and how many pages was 1,500 words? A week later I came back, essay in hand. I never showed it to anyone. The only people who read it were Judge Zoll and me. We spent half an hour reviewing it in his chambers and then he said, "This is very, very good. And I'm going to give you a break." And then his voice turned stern. "This is the only break you'll get from me in your life. And I don't want you to steal anything ever again. Because if you do, I'll hear about it." He warned me that if I did anything wrong, anything, he would know about it. He verbally kicked my butt. And I believed him.

I took my mother's car out a few more times, and I drank again, even though I was underage, but I never stole another thing, not even food, no matter how hungry I was. If I ever so much as thought about it, I heard the words that Judge Zoll had spoken. I could hear him saying, "I know where you are. Don't steal. Don't steal." Years later, when I began my U.S. Senate campaign, I bought a huge shopping cart of items at Staples. When I got to my truck, I found that there was a stapler buried at the bottom. I hadn't paid for it. And I thought of Judge Zoll. I walked back into the store; the clerks told me to keep it, but I insisted that they ring it up. I told them that I wasn't taking anything.

And after Judge Zoll's chambers, I cut off all of my long hair, for good.

Chapter Eight

BASKETBALL

We always knew that Wakefield was old. At the town boundary limits were signs that read, "Wakefield, founded 1644." But it began even earlier than that, in 1639, when the General Court gave a four-mile-square grant of land to the town of Lynn for a new village. Its first buildings were erected at a time when the only way to travel west was along the worn remnants of Indian paths. In 1644, the town had seven families living in seven houses, as well as a "humble church edifice," and it had taken the name of Reading.

Reading had no great role in the American Revolution, and the town was bitterly divided in 1812 over whether to side with James Madison and fight the British for insulting our sailors and our flag. The Old Parish, which would become Wakefield, was wildly supportive of Madison and ferociously against England, while the rest of the area was violently opposed to any war and also to President Madison. The passions ran so deep that supporters of the war were excluded from town offices, and within months the town broke apart, with the pro-Madison residents petitioning for their own charter for a new town, which they now called South Reading.

The town that they built was a place where the houses huddled close together, not quite as close as in Malden or Revere, which were older and nearer to Boston, but close enough, as if they were seeking warmth from each other when the winter sky turned gray and heavy and the sun set, enveloping the streets in darkness, by late afternoon. South Reading occupied one of the last bands of the old towns fanning out from Boston, each one marking a new notch in the western push of migration, of people laying claim to the virgin space beyond the crowded, noisy city. In the 1600s, the original town of Cambridge, across the water from Boston, stretched for thirty-five miles from the Charles River to the Merrimack and took an entire day's journey to cross.

South Reading was hardly as grand. But it was beautiful. Unlike the steep, sheer cliffs of Malden, South Reading was a collection of wavy hills and winding streets, of green, leafy trees and the remnants of woods where deer and other small game once roamed. The only wide-open vistas were down by the lake; the rest of the town rose and fell like the bumps of a weathered, prehistoric spine.

Its name was officially changed in 1868, twenty-four years after the Boston–Maine railroad extension was laid, when Cyrus Wakefield, who was the owner of the highly successful Wakefield Rattan Company and whose family had been residents of South Reading for generations, offered to donate the funds to build a new town hall. Residents assembled in a town meeting and decided to rename South Reading with "the new and significant name of Wakefield." The vote was taken on July 1, and the official renaming took place three days later, on July 4, a day of "excessive heat," accompanied by pealing bells, firing cannons, band concerts, a procession, and a special commemorative poem, which concluded:

No soft Italian scenes we boast,
Our summer skies less clear;
But prized the grandeur of our coast,
Our rocky hillsides dear.

No notes of foreign praise we swell,
Not, "Naples view, and rest!"
Our invitation is, "Come, dwell
In Wakefield, and be blest!"

The celebration was complete only after a full historical address, a grand celebration dinner held beneath a tent on the common, and an evening capped off by fireworks. It was still the same spot where, each Fourth of July, we watched fireworks and the town gathered for a concert and a parade. And we never thought of it as anything other than Wakefield. To us, that was what it was and what it had always been.

My personal geography of Wakefield was defined by the locations of blacktop courts and metal rims, by the places where I could play basketball. There was the court at J.J. Round, a strip of park one block down from the Oosterman Rest Home, or the basket up against the wall by the Franklin school, or Nasella field on Water Street. I would rotate among all the courts, looking for games like a sloop searching for a port of call. They were my destinations. In the summer, I would ride my bike up to J.J. Round Park, arrive at nine in the morning, and stay until nine at night, just hanging around, waiting for pickup games.

If I got hungry, I would go buy a slice of pizza. The parks were where the basketball stars would go, high school and junior high stars, like Bob and Jay Moore, and if I waited, I could play with them. "Hey, Brownie," they'd say, "you're on our team." I was good enough to hold my own on the court with them, and I always wanted to beat them. I would also go up and play with my friends who were my own age, like Bob Najarian, Bill Cole, the Gonnella brothers, Jim Healy, Bill Squires, and Don Flanagan.

If no one came, I would practice drills—quickness drills, wind sprints—and moves to improve my footwork like crossover running, crisscrossing my feet as I moved over the court. I practiced shuffling and speed. I threw rebounds, scooping the ball as it ricocheted off the backboard. I practiced boxing out opponents and form shooting. I'd start at the foul line and move around the entire box; when I had made the close shots, I took two steps back, and shot again in another ring formation. I'd finish with five foul shots. Every one had to be a swish shot, straight in the basket, no ball rolling around the rim.

I would ride there many days in the dead of winter, even when there was a foot of snow on the ground. I'd balance a shovel in my right hand or tie it to the rattrap on the back and steer my bike with my left hand, my ball tucked under my arm. I'd shovel the snow off the court, my breath puffing ice in the air. And then I'd stand on the blacktop and shoot. The ball wouldn't bounce in the cold—it died on the ground as soon as it dropped through the net—but I didn't care. I'd shoot for two or even three hours at a stretch, until my hands were so numb that I could no longer feel the pebbly surface at the tips of my fingers.

I began playing in leagues when I was still in elementary school. My first coach was Zach Boyages, Mr. Boyages, who ran the summer and winter youth basketball leagues. His sons, Mike and Ricky, played too, and he was known as Mr. Basketball; he had played and so had his brother. I spent hours over at his house, in front of his hoop, and he drove a group of us, six or seven kids piled into his station wagon, to invitational tournaments with other local leagues. He was grooming us like a farm team to move up to the seventh- and eighth-grade levels, where basketball was serious business. Mr. Boyages played a very important role in my life and that of many other young boys during those years in Wakefield. I still remember his warm smile and his love of the game.

Aside from football, basketball was the most competitive sport around in our section of Middlesex

County, with hockey and baseball close behind. The rivalries, like Wakefield–Melrose, Wakefield–Woburn, or Wakefield–Lexington, were fierce, taking on the sheen of Yankees–Red Sox, and the games were packed. This was what a lot of people did on a Friday night or a weekend: they headed for a gym and watched sweat-soaked teenage boys locked in battle.

That year, as I did every year, I looked at our schedule before the start of the season. I pasted it into my scrapbook, and I kept my own private score. I'd write down how many rebounds I got, how many points I scored, whether we won or lost, and how many points my opponent scored on me. I got so that I could look at the schedule at the start of the season and say, "We're going to win this one, this one, this one, and we'll really have to work hard at this one." Each week, I'd think about the game coming up and think about ways to be prepared.

Coaches motivated me by pushing my buttons, by driving me to be tougher, to work harder. We played a lot of our games in the high school, and I'd walk to and from there on Saturdays and Sundays, a good couple of miles each way, even when it was snowing. By eighth grade, we had scouting reports, and Coach Simpson would point out which one of our opponents was the high scorer or the best player on the team.

He'd say, "I don't know if we're good enough. I don't know if we have a chance to win. I guess if everyone comes and plays, we have a fighting chance." To me, each word was a challenge. I'd think: that guy from the other team isn't out on the courts shoveling snow and playing hoops in the winter; he isn't the one staying after, playing with the coaches one-on-one or two-on-two. It made me more determined to show him, to score the most points, to make my team better, to work harder. I'd get in the huddle and say, "We can do this, we can get it done." But there were a few times when I was younger, when we were down in the fourth quarter of a game in one of the leagues, and I said to the coach, "Just give me the ball." I was tall, I was quick, my favorite move was dribbling and driving straight to the basket, and I could be a scoring machine.

By eighth grade, though, we were a tight-knit team. Most of us had been playing together for years, in parks, in rec leagues, and with Mr. Boyages. In the chaos of becoming teenagers, on the court we had a sense of order. Basketball was something where we knew what the rules were; we knew what the time was, how to watch the clock, how long it took to tick off each quarter. More than school or family, basketball dominated our lives. We talked about the games, we sized up our opponents, and when they came to take us on, we treated it like war.

The kids from Woburn were the most daunting. A lot of them played football and were then recruited to play basketball because the team needed enforcers. They were guys with muscles, and we were scrawny little guys with narrow chicken arms. And we knew we would be playing these same guys right through high school and even into college. When we faced the Woburn Tanners, the Lexington Minutemen, or the Burlington Rockets, we knew that we would be repeating this same battle for years to come. We ran plays and we were competitive, diving for the ball and scrapping. I would dive for the ball and end up with scrapes on my knees and elbows, blood everywhere. Once, I said, "Coach, I'm thinking about wearing knee pads." And he said, "You know what, you can wear knee pads, sure. You don't want to get hurt. But do you know what the sign of a good player is? When you're out there looking around at all the different players, the guy with the scrapes on his knees is always the best player on his team. He is the one that we have to watch out for. All the ones who wear knee pads and elbow pads and mouthpieces, those are the people you don't need to worry about. It's the guys who don't care about their bodies—those are the kids you have to watch out for." So I wore my bloodied knees and elbows like a badge of honor, and I never feared diving.

Years later, I said the same thing to both my daughters. And when Ayla and Arianna played hoop, they always had bruised knees.

Before I went on to high school, Brad and Judy invited me to their August wedding. I came to the church and brought the entire eighth-grade team. When the ceremony was over, I took the team to the reception. Never mind that I hadn't been invited to the reception, or that the entire team hadn't been invited to the wedding. They had invited me alone, as the captain and one of their favorite kids, but I brought everyone along. I never knew that my coach and his new bride scrambled to add another table at the hall and that the caterers now had to feed a group of hungry teens. Afterward, all Brad said was, "When you get married, you owe us an invitation."

In high school I played freshman basketball for coach Bob Gesing—I wasn't allowed on the varsity team. But I never left the Simpsons. They lived near the high school and many nights, before or after basketball practice, I would drop by to visit, or for dinner, and Judy always managed to put together an extra plate of whatever she had been making. We'd talk about sports or how I was doing in school. Sometimes I'd tell them about Leeann. I wanted to be like Brad, so full of life and enthusiasm, so happy in his home with his beautiful wife. I was like a big, lost puppy loping around after them. That summer and the next I was asked to go to basketball camp in Maine, where all the Wakefield players went to hone their skills, and there was no money at home to pay for it. Then at the last second, miraculously, I was told that a slot had opened up, a scholarship slot. Except there was no such thing. Brad and Judy had paid out of their own pockets for me to attend. Later, when I learned about it, Judy called it "an investment in my future." I could repay them, she said, by sticking with sports and getting good grades. And that was all. They didn't want my money, she said. They wanted me to do well.

In high school, I had to play freshman hoop, where I averaged 25 or 30 points a game. Coach Gesing was a great coach and let us play exciting ball. But I kept asking him, "Can I move up? Can I move up and play?" And the varsity coaches kept saying that I wasn't ready. When we finally finished our season, the coach agreed to let me come up and play junior varsity, with the sophomores and juniors and a couple of seniors. A lot of the guys were players I had played with in the summer leagues. There was one game left in the season, against Melrose, one of our biggest rivals. I was sitting on the bench in the JV uniform, the only ninth-grader to come up and play in a very long time or perhaps ever. The first quarter went by, and I didn't play. The second quarter passed, and I didn't play. The third quarter went by, and I didn't play. Finally, the coach looked at me in the huddle and said, "Brown, you're in." He started me in the fourth quarter and we were down. I raced up the court and hit my first four or five shots in a row. I was taking hook shots from the foul line. I wasn't missing anything. And about halfway through, the coach said, "Get the ball to Scott." There I was in my first game, the youngest kid, the first time I had played with these guys on a school team, and the coach is saying, "Get the ball to Scott." We won, and it was about the happiest I had ever felt in my life.

I wasn't just a basketball guy. I had played baseball, each fall I ran cross-country, and in the spring I ran track. I still have a row of dark cinders in my knees from where I fell racing the 330-yard hurdles and missed clearing a couple. I got up and finished the race anyway, blood running down my shin. I threw the javelin and ran my events, the quarter mile, the half mile, the mile, and the 330-yard relay hurdles. I was a three-miler on the cross-country team, snaking up through the hills. My teammates, John Bowman, Brian Doherty, Rich Hansen, Bill Squires, and I had a lot of laughs when we were out running. Most of the kids on the squad weren't physically big, they weren't football or basketball or hockey players, but they were tough runners and good, nice kids. And in cross-country, we could be goofy, always cracking on each other, pranks that today would have gotten us suspended from school in less than a heartbeat. Sometimes, we'd sneak up on a guy running in the line, pull his shorts down, and race away. Or we'd rub Bengay arthritis ointment on his clothes in the locker room. I remember we pulled a prank on one guy, Billy Solomene, a hotshot freshman, when I was a junior or a senior. Back then, he was a scrawny little kid, but he grew up to be six foot three and one of the best triathletes in New England. Fortunately, he took it pretty well, and we still joke about it today.

I loved running. As a freshman, I won almost all the meets and all the invitationals. I ran varsity as a sophomore and as a junior, but then I quit. I thought I would try out for football. The coach hated me for that. I was the team's best runner as a junior, and he had an undefeated record for almost his entire career. But when the fall of my senior year came and I saw my cross-country buddies getting ready for our first meet of the season, all my friends lining up to wait for the signal, I changed my mind. I loved our meets, and this one was against Melrose. I hadn't practiced at all and hadn't run much during the summer. I just threw on a team uniform and raced for the starting line. I came in second on my team in the race and might have won if I had trained. I rejoined the team, and was a top runner by the end of the season, but the cross-country coach never forgave me for failing him in the beginning.

The biggest problem for me with cross-country was that it made me too skinny for basketball. When I was running, I was six foot one and barely 140 pounds. I was so skinny and worn-out that during a state meet in my junior year, I was one hundred yards from the finish line when I passed out cold. I was in the top ten of cross-country runners out of hundreds. But I was just too skinny, and this made me particularly vulnerable for the heavy muscle play of basketball.

And I lived for basketball.

In the years 1971–1972, there were twelve suspicious fires across Wakefield, at the movie theater and the Armor Fence Company, some empty houses, and even the industrial park that had housed Cyrus Wakefield's original rattan company. It got so that residents would wait for the wail of the fire alarm, wondering what would be the latest place to go up in flames. One of the worst fires, in December 1971, destroyed part of Wakefield High School, including the gymnasium. The students had to split classes, part of the school starting at the crack of dawn, and then leaving early, while the next round of students came in for the afternoon. Sports practices were pushed into the evening. But what I remember most was that the old high school had to rcbuild, and the town decided to turn that school into a junior high and then build a brand-new building away from downtown for the high school students—complete with a new gym and new basketball courts.

I remember walking into the unfinished gym when the poles and nets were up, but the final smooth, shiny composite floor had yet to be laid, and walking to each hoop, standing there and visualizing the shots from every angle, mentally seeing myself execute with the ball. I walked to the game hoops, the practice hoops, every last one, ringing both sides of the gym, and imagined myself there. When I got into the heat of a game, I could will my mind to remember just where I should be, ball in my hand, basket in my sights.

School in Wakefield was cliquey and divided, a little like the town itself, where the train lines marked off the East Side—with its blue-collar houses and Chevys and Dodges and Buicks parked alongside the curbs—from the more affluent West Side, where the doctors and lawyers and Boston bankers lived. There, the lawns were bigger and the owners hired lawn services to mow in the summertime, and it was not uncommon to see Cadillacs or Oldsmobiles in the driveways. Even within the two sides, there were divisions, the old-line New Englanders versus the newer immigrants. One section of the East Side was called "Guinea gulch," derisively stamped as an Italian ghetto even though it was just another ring of cross streets dotted with modest post-war, middle-class Cape and A-frame homes. Wakefield wasn't racially mixed, but it had its own ethnic or class tensions and put-downs, and they carried over from the homes to the hallways to the gymnasium to the playground. By ninth grade in Wakefield High, kids separated into factions. There were the jocks or heroes and the freaks and the others, and they traveled in packs and marked their territory as thoroughly as dogs. I was a jock, but I was also a loner, one of the kids who wore the same clothes over and over, and I wasn't afraid to take anybody on.

During my freshman year of high school, I was shooting around with Jim Albanese and Billy Cole at Nasella Field on Water Street. It was on the East Side of the tracks, a place with a baseball sandlot, a soccer field, and basketball courts, surrounded by trees, where the punks hung out. It was a warm early evening, a men's softball league was playing on the diamond at the far end of the field, and I was back near the roadside, a basketball in my hand. I turned just as one of the Fotino brothers and his crew sauntered in. There was a group of them, maybe eight to ten, the two brothers and their friends. They traveled in packs, cultivating the look of street guys, tight T-shirts with cigarette packs rolled up in the sleeve of one arm. One of the Fotino brothers kicked the basketball. I retrieved it and continued to shoot. And he grabbed for it again. I got it, and yet again, he lunged. He was older than me and bigger. We had fought once before, a knock-down, drag-out fight that I had ultimately won. He was back now, looking to even the score.

He came at me with a mocking face, a cigarette dangling from his mouth, backed up by his friends. Anticipating that he would want to throw the first punch and thus gain the advantage, I pulled my fist back and let it fly, straight at his mouth, the cigarette collapsing between my knuckles and his teeth like an accordion. Then we hit the ground, the hard, hot blacktop. We rolled like animals, each clawing for the kill. My elbows were bloody from the hot black tar; my knees were on fire, dragged and shredded; but I wouldn't let go. He knocked my arm and kneed me in the stomach, and then I reached around with my left arm, my shooting arm, and grabbed him in a headlock. My friends were already racing across the field to get some of the men from the softball team. Joey's friends were kicking me, driving their feet into my back, aiming for my kidneys, or trying to kick the soft underbelly of my abdomen. And I was slamming his head against the ground. "Every time you kick me," I spat out, "I'm slamming his head." They kicked again, and I smashed his head, boom. Kick, boom. Kick, boom. I took whatever punishment they gave me and I gave it to him. There was no way for them to know that I had long ago learned how to take a beating. Then the men from the softball team ran up and separated us, and it was over. We were pulled apart: he was howling and spitting, and I was shaking with adrenaline and wiping the blood from my hands. It was done. When our paths crossed, there was now an established order of things. Unless he could find me alone and he was with all his friends, he knew to stay away from me. He would not get another chance with me, one-on-one, again.

I wanted them to know that no one could mess with me. If they came looking for trouble, I would give it back to them. The problem was that trouble would come looking for me. Even after that summer, it kept coming.

My high school coach was a legend, a man named Ellis "Sonny" Lane, who lived in Reading. He came to the school in 1970 from Stoneham, and he was in his own right an athlete of distinction—a Middlesex League all-star in basketball and baseball. At the start of his senior year, he tried out for the football team for the first time and was instantly made starting quarterback. He played semipro baseball and won a four-year baseball scholarship to college. But now his turf was basketball. Wakefield was one of the smallest schools in the Middlesex League, but that didn't matter to Coach Lane. Winning did. He wanted us, he willed us, to win. This was an era when coaches wore jackets and ties to all the games, when there was a press box right off the court in the high school gymnasium, when high school games were chronicled with almost the same obsession as the Celtics. Sonny Lane showed up for our games with his assistant coaches in a large, plaid sport jacket with mile-wide lapels and a bold, oversize pattern that verged on blinding. But that was where the flash ended. He didn't like any flash in his games.

Coach Lane didn't swear, but he would say, "What the hell was that?" On his court, there were no behind-the-back dribbles, no dipsy-do shots—the plays I had been working on for years. It was fundamental basketball, while I leaned toward the flamboyant. My heroes were John Havlicek, the great Celtics player who made his career running on the court as a fast-break star and a clutch stealer and outside shooter, and "Pistol" Pete Maravich, the scoring machine, whose repertoire was ball tricks, behind-the-back passes, head fakes, and long-range shots. I read their books and watched their films. I wanted to be just like them. But Coach Lane wasn't interested in would-be Havliceks and Maraviches on his team. His philosophy was that we didn't set up plays for individuals. Everything was about the team: it was pass first and shoot second. He was the type of coach to say, "Well, hopefully you'll have a good game, but what about the other kids on the team? How are you going to help them? It's not about you or any single player." He wanted us to be a team.

If I did well, he'd say, "Oh, good play," but whenever I screwed up, he'd unload. "Hey, Brown, you know that move you just did? Take it to bed with you. I don't want to see that ever again." His favorite line was, "Brown, what the hell were you thinking? Brown, take that shot to bed." He'd stop practice and mockingly say, "Oh my God, Brown, that move is going to get you in the Hall of Fame. You keep living on that. You keep living on that shot and you'll be sitting on the bench right next to me for the rest of the game and for the rest of the year, Scott."

And back then, I worried that this was exactly what would happen. My sophomore year was frustrating. I had a great start to the season, but then I sprained my ankle in practice and I got a blister on the bottom of my foot, which developed a staph infection. It was absolute torture when I shoved my feet in my sneakers and ran. I had come in as a high scorer, but now the coach wanted me to play defense because the team already had a lot of scorers. He had me run defensive shuffles around the gym to work on my footwork. He had me dog the other team's best players, and was constantly saying, "Scott, sacrifice your scoring for the team." I did. I sacrificed everything. I got to the gym first for practice and stayed after everyone was gone. I learned to pass the ball, to assist, to shadow the other team's players. But in the big games, I still scored: 17 points against our archrival, Lexington; 12 points against Belmont. The newspapers called me "Super Sophomore." That summer, I got a shooting glove to train my hand to keep the ball in my fingertips. I became a perimeter shooter, before there was such a thing as the three-point line.

When I came back to practice, Coach Lane was every bit as much of a hard-ass.

We got scouting reports on all the teams we played and watched films of them before games, analyzed them, were tested on them, and then watched films of ourselves, until after seeing so many black-and-white images and dissecting all my imperfections, I was ready to quit or go home crying. And so were most of the other players on the team. It didn't matter how good I was—Coach Lane was riding me all the time, until I believed that I sucked, that I was a fraud, that I was nothing running up and down the court, driving to the edge of the foul line. Finally, I confronted him. I said, "Coach, you're yelling at me all the time. What's up? I can't be that bad."

He shocked me by saying, "Listen, you're not. You're great. You're a great player. You're a hard worker. You're the hardest worker on the team, but if I don't yell at you, then I can't yell at the other kids. When they see me yelling at you, they say, 'Oh my God, he's all over Scott.' So when I yell at you and then I yell at them a couple of times, they don't take it as bad." He went on to say that some of the other players can't be yelled at because they'll go home crying to their parents. "I can yell at you," he said, "because I know you'll just get angrier and angrier and work harder and harder."

And he was right. He was right too because I would

never think to go to either of my parents and complain. By now I had much bigger problems at home. My dad had moved out to Western Massachusetts with his third wife, and my mother was remarried again, for a fourth time.

"When I stop yelling, you start worrying, OK?" Coach Lane added.

"OK, coach," I said.

Chapter Nine

LARRY

Around 1999, a house came up for sale on June Circle in Wakefield. It was a brown house of deep red brick, with colored shutters and a red front door, and for a few seconds, I contemplated buying it and then burning it down.

I helped build that house, at the end of the cul-de-sac, up the steep hill from the high school. It was the last lot on the street, backing up to a small woods. Larry McShane owned the lot. He was a friend of my mother's from Wakefield High, and he was the baseball coach. He was the man who taught me how to throw a knuckleball and how to throw a fast pitch. I had a couple of good baseball seasons, as a lefty pitcher, throwing fast and mostly wild. He taught me how to drive too, although I didn't mention that I didn't really need lessons, but he did teach me how to coast up to a red light, to go light on the brake and let the car glide. "Anticipate the stop sign," he would say; "keep it smooth." And he taught me about horse racing and the racetrack.

Larry, my mom, Leeann, and I used to go up to Salem, New Hampshire, to the track in Rockingham Park and watch the Thoroughbred races from the sprawling, raised grandstand above the dirt. Larry liked to go when it was sunny, when you could feel the heat from the sky sinking deep into your bones. I was Larry's runner, from our seats to the betting window. He taught me how to read the race cards, which horses to bet on, and which to avoid. My job was to sprint with the card to the window, stand in line during the countdown of seconds to the starting pistol, and buy the tickets before the gates clanged open. When Larry won his bets, he would send me back to the windows to collect his winnings, and I got to keep 10 percent. He was a good bettor, a strategic one, careful and meticulous with his money, and there were a few afternoons when I'd walk away with $100 in my pocket, while Larry counted out $900 or $1,000. To this day, I can read between the lines on a race card because of him.

Larry hired me and sometimes a couple of my friends to clear out his lot. I cut brush and carried it to the mulcher and cleared away rocks that were sticking out of the ground. I was strong and not afraid of working up a good sweat, of banging away with a shovel or a pickax. One summer when my mom was on welfare, I was eligible for the U.S. government's Comprehensive Employment and Training Act (CETA) program, which provided jobs for low-income students. I did rugged work outside, painting fire hydrants and town benches, among other things. Another summer, the Wakefield Municipal Gas and Light Department hired me as a ditchdigger, burrowing into the ground so the department could lay cable or bury lines. At the June Circle lot, I hauled rocks and brush and cleared off the ledge so that workers could blast it big and wide enough to dig a cellar hole and lay the foundation. It took about a year to build the house. It was reddish brown brick on the bottom, with a brown-painted clapboard second floor. In the basement, Larry built a wet bar from one of the trees on the lot that had been cut down. I helped him sand it and stain it. And over the garage, he put in an in-law apartment for his mom.

Larry had time to build a house, coach baseball, and go to the track because he did not have to work. He had been injured in an industrial explosion. The skin on his face and much of his upper body had been burned by heat and flame and had healed into scars. But what had borne the brunt of the blast's force were his hands. There, the skin was a knotted mess of scars. The tips of his fingers, down to his knuckles, were gone, amputated in a single boom. When he opened his hands, there was nothing there but palms and stubs. After the accident, he received a big cash settlement, enough so that he did not have to work again. But inside, he was angry. And as each year passed, the anger festered. He was a solidly built man with a thick Boston accent; he dropped his *r*'s and hardened his *a*'s. He compensated for his physical flaws by being charming and witty, but that was largely for public consumption. In private, he was a different man.

Not long after Larry moved into his new house, he married my mom. They had, I think, become reacquainted at a local baseball game, perhaps even one where I or one of my friends was playing. They had known each other in high school, my mother the pretty cheerleader and Larry the tough kid, in your face and a bit of a bruiser, skating at the edge of trouble even then.

He married her and she married him even after Leeann and I had heard and seen them in the parking lot of our latest apartment, a cluster of squat garden apartments closer in to downtown. It was after a bitter cold snow; the plows had come through the parking lot, scraping and pushing the fallen flakes into high, slick mounds. We heard them from inside, upstairs, through the closed windows. They were screaming at each other, and then Larry began to hit my mom, beating her with his stub-fingered hands right up against a car in the parking lot. Leeann and I ran to them, shouting and screaming, adding our voices to the whacks and thuds and to the rising din. I pulled them apart. Leeann was crying. My mother was crying, and Larry was seething. But the next day, he apologized. He showed up on our doorstep with a fuzzy kitten and an armload of presents, charming and sweet-talking my mother, and within a couple of months, she was Judy McShane.

From New Hampshire, Gram and Gramps asked her, "Are you sure?" before the wedding, but beyond that, they did not interfere, and so she made her choice again. We packed up and moved to June Circle, with its faux Colonial plaster walls and exposed beams and wallpapered bedrooms, a place more horrible than the house of *The Exorcist*, which I watched on the towering screen at the local drive-in.

In the beginning, the fighting was just between Larry and my mom. I stayed out of the way. To me, battling was what married people did; fighting came with the new house and the wedding ring. But Larry also drank and my mom periodically drank. That changed everything. Sometime after those first few months, their fights turned vicious. And alcohol lubricated nearly every one.

Because he didn't have a job, Larry was around the house much of the time. He liked puttering around and doing yard work, and he liked coaching baseball. He taught many of my friends, including my buddy Mike Quinn and also Bobby Moore, the intricacies of the game in the big diamond across from the high school gym and the running track. But there was always something sharp-edged about him, something coarse and rude. His moods and his emotions could shift without warning. I got to be as adept at reading Larry as a farmer is at reading the weather, season after season, scanning the horizon for the first signs of whatever new front is coming. I would come through the door and ask him, "Hey, how ya doing? What are you up to?" and I could tell just by looking, by the tilt of his head, just what kind of an evening was waiting. Sometimes he could be pleasant, but then the simplest question would be met by "Why the fuck do you care? I'm doing whatever I'm doing." I would say, "Oh, OK," and then retreat to my room or some other part of the house or even head back out to a basketball court or to a friend's. If I got as close as three or four feet, I could smell the alcohol, the bitter, wafting scent of scotch or beer puffing out in small clouds with his breath. And he would look straight at me, with a gleam in his eye, as if to say, "Let's play. Here we go." His jaw would clench, the scars drawing tight against his skin, and he would begin to light into me.

Living in that brown-and-red Colonial house was like walking on eggshells all the time. I couldn't turn the stereo up, couldn't come home late, couldn't eat certain things. Each week, it seemed there was some new "Larry rule," something else that I couldn't do. It was a combination of fear and intimidation, the way sometimes he would sit in his home office, counting his bonds. He would take out the papers and say, "This one's worth $50,000, this one is worth $100,000," trying to impress me and knowing that most weeks, I barely had $10 to my name. I was babysitting for Coach Lane and others, knocking on doors to mow lawns, anything to make some extra cash, while he was at home, counting his stacks of bonds.

He chipped away slowly, layer upon layer, until we were all left terrified in his home, in the house whose land I had helped clear, whose walls I had helped to paint, whose wet bar I had helped carry in with my own hands. It was Dan Sullivan again, but it was Dan Sullivan night after night, Dan Sullivan able to inflict the most searing kind of pain. Larry wasn't just a brawler, blindly lashing out, flailing away, lucky to hit something; he was a planner, a street fighter. He knew the weaknesses and vulnerabilities of his opponents as well as he knew the horses on the race card. He studied and he read them.

It was at night mostly when the confrontations came. I would be in my room when I would hear the first sounds. A door would slam; there would be banging, pounding, and yelling. Then the screaming and the crashing. It was always in the middle of the hallway, upstairs, second floor, right in front of the brown tile bathroom. And he would be going after my mom. She would lock herself behind a door, and I would come charging out of my room, maybe twenty feet away. And Larry would be waiting. He was five foot ten, but he was strong, with a thick neck and muscled torso. He seemed to delight in taking me on. We would grapple, wrestle, bang against each other, throw each other up against the wall, slam a head, a shoulder, an elbow to see who would wince first from the pain. And then he would uncoil his most lethal weapon, his hands.

He would jab at my eyes, my face, my neck, my

stomach, my ribs, ramming me with those spare little points, the sawed-off stumps of his fingers that were nothing more than skin-covered bone. He never had to make a fist; the remains of his fingers might have had almost no sensation, but they were brutal in their efficiency, in their ability to maim and bruise deep beneath the surface of the skin.

Sometimes he was in the hallway, but other times he had wedged his way into the bathroom with my mom. The door would be locked, and I'd be banging, daring him to come out, anything to get him away from her and to take me on. Suddenly, the door would swing open, and he would be right in my face, now daring me, arms on me, grabbing me by the throat, pushing and pounding. And as he did it, he'd say, "You'll still have to go to sleep tonight. When this is over, you'll have to go to sleep. And when you're asleep, I'll take you out. That's when I'll come in."

In that house, amid my poster-covered walls, I became a chronically light sleeper, listening for the first telltale bang or slam, listening for the click of the doorknob or Larry's feet. I started sleeping with my door locked and a heavy wooden baseball bat tucked away in my room, for protection. And I slept with my basketball.

My mother, when we fought when I was younger,

had repeatedly threatened to break my sports trophies. She knew the quickest way to get me to back down was to head for the hard-won basketball, baseball, and track awards that I kept in an ever-expanding display in my room. Larry threatened to break something else: my hands.

He told me he would break my hands so that I couldn't shoot, so that my basketball career would be ruined. He could do it too, with a knife, a bat, a metal shovel from the garage, or just the full force of his body against my own. That was the threat he held over me, all those years, the threat of hurting my hands. Because I knew that if I couldn't play basketball, I would never, ever get out of there. My ability to shoot was my only hope, my only plan of salvation.

By high school, I realized that I was good enough to get a scholarship, and a scholarship would take me away to college. Without a scholarship, without that money, I would have no chance to go on to anything. And I knew too that I had to get out. Every time Larry fought with my mother—every time he backed her into a room, and I had to come to the rescue, with Leeann standing in the hall in her little nightgown, screaming and crying, and my mother holding her back, away from the two males locked in mortal combat outside that bathroom—every time, I was afraid that as I got bigger, as I got stronger and he got slightly older, as I got quicker, during the next fight he might employ his street smarts and find something else that could be used as a weapon. Or that I might lose all control and really hurt him, or worse. When we fought, I held back. We would never fight to anything other than a draw.

Larry liked to taunt me when we fought, "Come on, come on, hit me, you chickenshit kid. I know the cops. They are friends of mine. What do you think the headline will read? 'High School Basketball Star Arrested for Hitting Stepfather'? The cops are friends of mine." And I thought he was telling the truth. I didn't realize that he didn't know many cops, or that the ones he did know hated him. I was certain that if I ever really hurt him, I would go to jail, and everything that I had worked for would be ruined. That was always in the back of my mind, and it scared me too, every bit as much as his vow to break my hands.

The worst of it was, when it was over, when the alcohol or the hangover had worn off and life had resumed at June Circle, oftentimes my mother would pretend that nothing had happened. She'd ignore it, the way she ignored the marks on her skin or on mine. Sometimes, she'd say, "I'm sorry," or, "We shouldn't have to go through this," or, "I'll take care of it." At times, I almost believed her, but then nothing would change. Larry could make all the promises he wanted, but not one of them mattered after he had a few belts of scotch or a vodka tonic in his hands. You could smell the fear in our house. We never quite knew when Larry would blow, and we were consumed with the waiting.

I understand now why my mother stayed. I understand that she was a woman who had scraped by on and off welfare, in tough, low-paying jobs. I know that she thought she needed the security and stability of a man, that she had no real profession, no true identity of her own, that nearly all her friends were married and mostly happy and she wanted to be like them. I understand every rationalization, all the bargains she made with herself, the excuses, the compromises. And everyone else looked the other way. My dad was in Western Massachusetts. I rarely saw him. My grandparents were in New Hampshire, growing old. I wasn't going to call them. People knew that Larry was a jerk, that he went around bad-mouthing me. My teammates and my coaches saw me arrive wired at practices or at games after our confrontations. They knew small bits of what was happening from what I said to them. Larry liked to go after me around game time or track and field time. He liked to rattle me before a meet, a practice, a date, or a game. He knew my schedule, and he planned his battles. But no one intervened. Even Larry's mother, who lived with us in the apartment over the garage and heard everything, ventured out only once or twice to implore him and my mother to calm down. Later, when the police came, all that happened was that both sides were encouraged to cool off, as if that would be the end. And my mother and Larry would take a temporary breather, while I waited for it to start up again.

I took it out on the court. Every bit of anger, I unleashed on the varnished floors. I ran faster, I worked harder, I rebounded better. I'd be more aggressive on defense. The court and one inflatable ball were going to take me out of here. That was my promise to myself, day after day. Some summers I played as many as 180 games, 3 to 4 games a day, five to seven days a week, 17 to 28 games a week for the entire summer. Plus the drills. I scraped together the cash for mini camps too—camps like Dave Cowens, John Havlicek, Nelson/Sanders, Calvin Murphy, and New England Basketball, often going with my teammates. These were the places where the best players in New England came together each summer. There was nothing I didn't do or wouldn't do to work on my game.

Coach Lane saw it. He'd say, "Whatever you got—

Brownie, whatever you're thinking, you just keep thinking that. You just keep working. Love it." And he'd ask, "Why doesn't everybody be like Brown over there? Hustling, diving, scrapping?" But those kids didn't have what I did. They didn't have Dan Sullivan or even a bloody playground fight with the Fotinos to prepare them for the battles in Larry's house. They didn't have the accumulated weight of dozens of junior high games when I would be scanning the stands for my dad's face, wondering if he was actually coming. When I thought he might show up, I never had my best game because I was so worried about whether he would actually be there. And the couple of times he did come, I overplayed trying to impress him and I'd get out of my groove. But gradually, I taught myself to tune it out, to ignore everything.

And these other kids on the team weren't breaking into school just to shoot for a few hours in the gym.

For years, from junior high on, when there were snow days, I'd hitchhike or walk a mile and a half to the high school or one of the other schools and walk around the school building, trying each window, pushing against the rims until I found a loose one, a place where someone had forgotten to twist the lock. I'd pry the window open, shimmy in over the sill, and head for the gym. There, it was warm and I could shoot for hours in the quiet of the room. I'd practice my whole routine of drills, foul shots, getting the ball with the basket behind me, then fake right, shoot left, and put it in the basket off the backboard. I taught myself to score from the corners, even how to shoot when I dropped to my knees. Eventually, after one of the coaches found me, he gave me a key, so that I could get in through the regular school doors. "Just lock up when you leave," he said.

Before each game, I had a ritual. I would listen to music full blast, David Bowie, Queen, and Aerosmith. Bowie and Aerosmith were my favorites before games. I'd always wear a certain kind of clothes, certain socks, certain sneakers, until I had a bad game and convinced myself that they didn't work and that I needed different ones. On my feet, I wore red Converse high-tops. Everyone else wore white sneakers, but I wore red, and I wore them so hard and so long that they had holes all the way down to the rubber. To keep them going, I'd put pads inside and tape them up with bands of sticky white athletic tape or gray duct tape. And in the gym, before every game, I'd stand in a corner and warm up with the ball. I'd revolve it around my body, behind my back, through my legs, passing it from right to left, left to right, until it was a blur in my hands. I'd dribble it around and under, rise up to shoot, and pass it from hand to hand. I wanted to feel the ball until it was almost an organic part of my body, until every move I made could be done with a basketball in my hands. I'd visualize myself at the basket, where I'd stand, the precise moment when I'd pull my wrist back and let it flex forward with just the right amount of spring. I'd go over the scouting report one more time. I would be thinking: Who do I have to guard? Is he lefty? Is he righty? How does he play? What does he do? I stood in that corner and I thought a couple of moves ahead.

My coaches taught me to do that, and any team that played us knew that they were in for a real game. We were a very physical team; we excelled at the pick-and-roll, blocking the other team's guard and then breaking away to pivot toward the basket so that our players had a clear shot. We were ferocious on defense. The moment the ball went up, we were trained to immediately find our man on the opposing team and box him out. We ran the traps and we ran our plays, but no matter how well prepared we were, there was always a jolt of anxiety running around the gym before each and every game. We were always wondering: How could we beat this team? I thrived on that, on the burst of adrenaline.

From the moment I stepped onto the court, I could hear everything. I heard the fans, my girlfriend, the jerks in the bleachers screaming for the opposing team, the other players, the kids on the bench, my coaches. I could hear almost each and every word as if it were spoken in isolation. Even in high school, I would look to see who had come—my girlfriend, my mom, Coach Lane's wife, or maybe Brad or Judy Simpson, occasionally my dad. Were any of them proud of me? Were any of them watching? But my focus always had to be first on the game. I could tune out the trash-talking from the stands, but the trash talk from players on the court only increased my motivation.

And then there were those near-perfect games, when the difficult shots, the lucky shots, the first shots, everything just went in, when the ball arced and sank effortlessly into the basket, and the plays came together with incredible speed, players passing, feeding the ball off each other; that was when we hit the zone. At those moments, we just didn't miss, we didn't foul out, every fundamental was strong, and we won.

And what I could tune out most of all on the court was home.

I lived in two different worlds, mine and Larry's. Most weeks, I spent as little time as possible in his house. I showed up basically to sleep, eat, and change. The few times I went there with my friends to hang out by the pool or in the basement with the bar, it escalated into a verbal confrontation. There was always an issue: we were too loud, we shouldn't be in the pool, or Larry would yell that someone had taken his beer or, more important, his booze. And then there was the risk that it could spin out of control into some larger clash of shoving, wrestling, and those lethal half-fingered hands. So we simply stopped coming around.

My friends were all athletes: Mike Quinn was the captain of the football team. Bruce Cerullo was the captain of the wrestling team. Dave Turner wrestled as well. Jimmy Healy, Bob Najarian, Billy Cole, and Mark Gonnella played basketball with me. Mark Simeola played on the soccer team, and so did Paul Seabury. Bobby Rose played football. We were good students too, and we were very competitive, with everyone else in the school and with each other. We ran for the same school offices and tried to best each other at whatever sports we played. It was a friendship and a rivalry that created a powerful camaraderie and a deep bond, which still stands today. We would have given our lives for each other. One of us in essence did. After college, Paul, who is diabetic, learned that his kidneys were failing. Mark, without a second's hesitation, offered to donate one of his kidneys so that Paul might have a rich and full life. I am thankful to be counted among these friends.

As a group, it was a lot of the same kids that I'd been

hanging out with since I was eleven or twelve, and very little about our routine had changed. We'd go to someone's house, someone with a tolerant mom, and eat through the fridge before we went out for the evening. We might go play miniature golf down on Route 1 or go to a drive-in. Sometimes we went to the youth center, where there would be a crowd hanging out playing basketball. And sometimes we walked with our girlfriends over to the woods at the edge of the cemetery, where we made out in view of the staid, carved marble headstones.

One of the few nights when we were at my house, and Larry must have been out or something, Mike Quinn brought over a dirty movie. We told my mother that we were going down to the basement to watch football films to help Mike prepare for the big game. She called down, "Let me see," and we answered, "No, no, they're highly classified. Coach doesn't even know we have them. So we'll just finish up. We know how to work the projector." And for years afterward, the guys would say, "Hey, Brownie, does your mom still want to watch the classified football film?"

We had parties too with beer that we bought, because back then liquor stores didn't card anyone. But we were just silly, never crazy or hurtful. One time, right before my mother married Larry, I had about thirty kids over to our apartment on Salem Street, dancing and messing around, when my mother wasn't at home. We were so loud that the tenant downstairs began banging on the ceiling with a broom and threatened to call the cops. In minutes, we had bundled everything up in trash bags and gotten the room spotless, and I sent almost everyone packing. When the cops rapped on the door and said they had a complaint from someone else in the home, there were only three of us there watching a baseball game. I said, "Oh geez, I'm sorry. We have been loud. It's a great game." And they came in and looked around and saw nothing.

I had two more run-ins with the cops. One happened when I dropped by a wild party in one of the really big historic homes perched at the edge of Wakefield's lake. Kids were hanging from the chandeliers, which had come out of the ceiling. There were over a hundred kids, most were drunk, and just after I arrived, the cops came. Everyone else bolted and the cops hauled me down to the station along with about five other guys, and made us all pay restitution. I tried to argue that I hadn't done anything, that I'd barely been there. And their answer was simple: "Well, you were there. You should have known better. You should have left." So I ended up paying a couple of hundred dollars that I didn't even have in restitution.

But the second incident was even more meaningful. One time, I was the designated driver as we were tooling around town, with two or three cases on the floor of the car. We always had a designated driver. We were not stupid enough to combine driving and drinking, but my friends were all drinking. And then a cop car pulled up, flashed its lights, and pulled us over, and the officer looked in. He looked up and said, "Guys, you can't do this. This is not right. And you know it's not right. You guys are the young leaders of this community and you shouldn't be doing this." He made us take the beers—probably two cases, forty-eight beers—and dump them. Crack it and pour, crack it and pour, forty-eight times, until it was done. Then he said, "You know what, Scott? You take each and every one of these guys home and when you're done, you call me and let me know that you took them home." So I brought them all home and called the officer. And it was done. It was a time and a place where cops could do that, where, like Judge Zoll, you could get a second chance. And we didn't do it again.

I also had another person determined to keep me out of trouble: Coach Lane. Once or twice a week, often on the weekends, he would ask me to come and babysit for his three children, two girls and a boy. I'd watch TV with them and play a little catch or hoop in the driveway. Coach lived in a comfortable split-level home, with a kitchen, living room, dining room, and den downstairs, and three or four bedrooms upstairs. But it was so far removed from everything I had known.

The babysitting gave me some money and it gave me a chance to eat better or take a girl out and treat her on a date. When I stepped into his house, I remember thinking, Wow, I get to eat whatever I want, and I'm getting paid. And it probably cost the Lanes a heck of a lot more in food than it did in babysitting fees. I'd eat anything, popcorn, chips, cold cuts, pizza. I'd eat jars of pickles. It wasn't that there wasn't food at Larry's house, and my mom did try to cook, but at this point, I couldn't bear to be home.

Just turning onto the small cul-de-sac of June Circle was stressful. The moment I walked through the door, all I could think was, "What dilemma is there going to be now?" I was constantly watching, wondering what was next, where the next big problem was going to come from. I would be thinking, "Is everybody all right? Is everything all right?" Even simple stuff, like: Are my trophies OK? Did anyone take my stuff, or wreck my stuff? I would worry whether Leeann was all right or whether my mom was. I would think: Is my mom drinking again? Is Larry drinking? Is Leeann getting in trouble with her friends?

While my friends worried about their grades, about whether their parents would let them borrow the car, or about fights with their girlfriends, I was thinking: Is tonight going to be the night when Larry finally breaks my hands?

My first basketball photo, surrounded by my father's trophies.

Me, about age one, around the time my parents separated.

In my Webelos uniform with my sister Leeann.

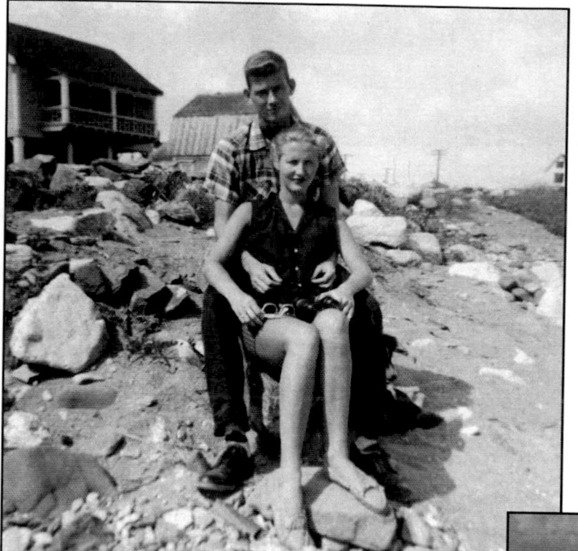

My parents, C. Bruce Brown and Judy Rugg, about the time they married, in Rye Beach, New Hampshire.

My mother, Judy, in June of 1967.

My father, Bruce, cooking in his kitchen. We have almost no photos of us when I was young.

My grandparents Philip and Bertha Rugg, who were a constant source of love and inspiration to me.

Me with my grandfather. By the time I was in college, I towered over him.

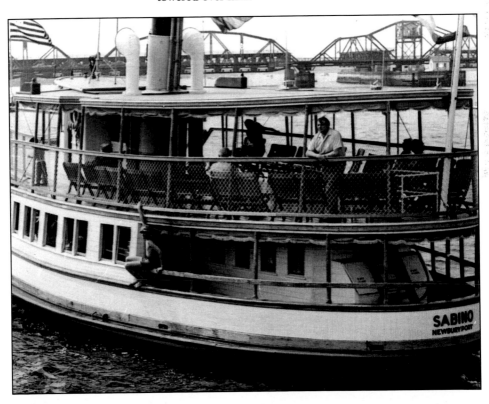

My father's coal-fired boat, the *Sabino*, on the water around Newburyport. I am the skinny kid sitting on the lower rail. I was about thirteen.

One of my summer junior league basketball teams. Many of my teammates have remained life-long friends. I am in the top row, on the right, holding two trophies.

Out-maneuvering an opponent during a hard-fought summer-league game at Dom Savio.

I was known on the court for speed and scoring.

Basketball was my ticket to Tufts University, where by my senior year I had scored nearly one thousand points in four seasons and had become cocaptain.

TUFTS 1977-78 BASKETBALL TEAM
Record: 16-8
Manhattanville Tournament Champions
Ranked 14th in National Div. III Poll

Front Row, left to right: Benji Williams, Ken Walker, Mike Rubin, Co-Capt. Mark Craigwell, Co-Capt. Bill Gorra, Tim Skaggs, Alvin Whitley and Admin. Asst. Matt Kaufman.
Second Row: Asst. Coach Bill Endicott, Jim Campbell, Paul Garrity, Ron Woods, Tom Manning, Kevin O'Brien, John Caragiorgis, Scott Brown and Head Coach John White.

My freshman-year team. I am seated next to Coach John White. Many of my teammates went on to highly successful careers.

COSMOPOLITAN

Helen Gurley Brown, Editor · 224 West 57th Street, New York, New York, 10019, (212) 262-7916

April 29, 1982

Mr. Scott Brown
1082 Main Street
Melrose, Massachusetts 02176

Dear Scott:

Congratulations! We are so pleased to tell you that you have been selected as Winner in Cosmo's Centerfold Contest. You charmed us all on your recent visit to New York and we look forward to seeing you again. We hope you enjoy seeing yourself in the June issue of Cosmo—we know our readers will!

All our best wishes,

Helen Gurley Brown

P.S. Please call Janice Trama at 212-262-7892 with your social security number and we will put through your check for $1000.

COSMOPOLITAN IS A PUBLICATION OF HEARST MAGAZINES, A DIVISION OF THE HEARST CORPORATION

My letter from Helen Gurley Brown announcing that I had won *Cosmopolitan* magazine's centerfold contest. My sister Leeann sent in my nomination.

Egon von Furstenberg
800 Madison Avenue,
New York, New York, 10022 (212) 737-8877
London • Milan • Tokyo • Hong Kong • Mexico City

A high-fashion ad for Egon von Furstenberg.

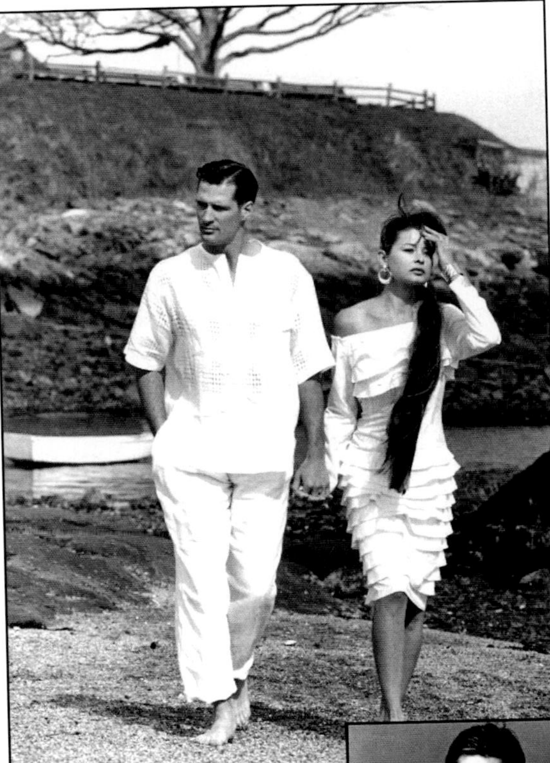

Gail and I pose during our first modeling assignment together. Away from the camera, we had eyes only for each other.

My modeling card is a complete 1980s flashback.

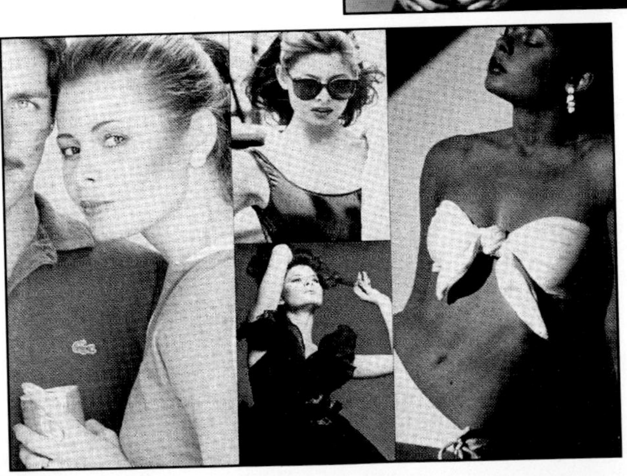

Gail's modeling card.

I was incredibly proud to join the National Guard and to serve.

Taking a break during infantry basic training at Fort Benning, Georgia.

On our wedding day, with my parents. We did not do a full family photo with all the relatives. There were too many conflicts.

I am grateful that my grandparents lived to see me happily married, with a family of my own.

One of many moments of happiness with my new bride.

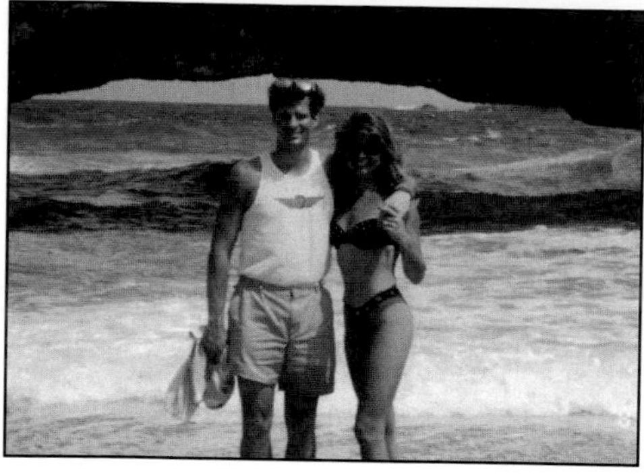

In Aruba on our honeymoon.

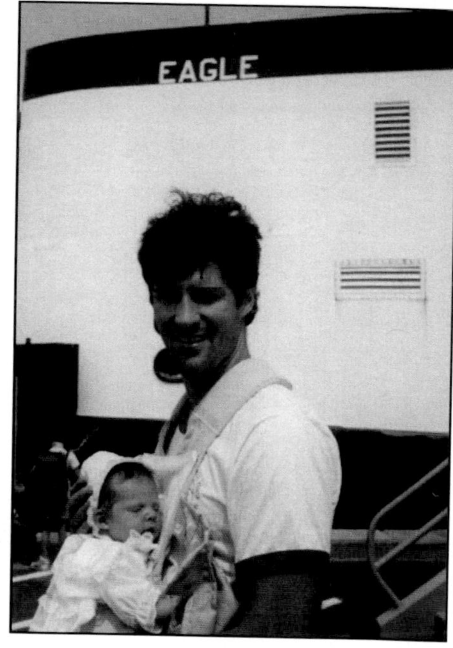

Ayla in a Snugli. I was overjoyed to be a dad.

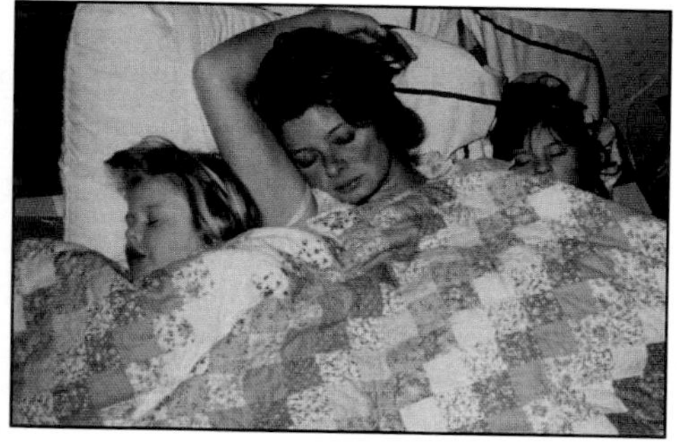

Gail's morning news job had her out of the house before 3 a.m. Sleep for all my girls was precious.

Our happy family, so different than what Gail or I had known.

A beach holiday after I had completed a triathlon. Ayla isn't quite tall enough to get the rabbit ears over Dad.

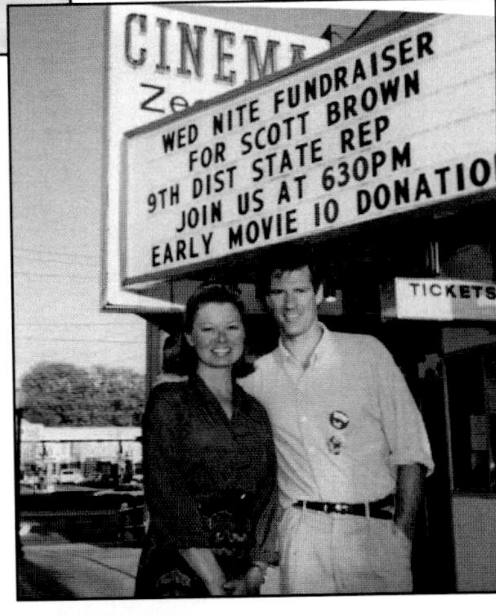

Gail encouraged me to run for office, and her personal support has been crucial to my success.

My first day in Boston at the State House, being sworn in as a Massachusetts state representative.

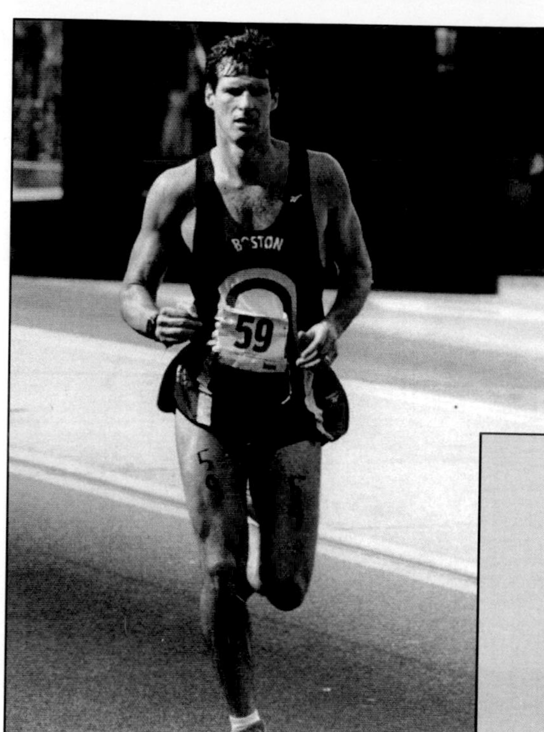

I love competition. This was taken at the National Duathlon Championship in Ohio back in 1995.

Finishing the first leg of a triathlon, the swim, in 2010.

Crossing the finish line in 2010. When I run, bike, or swim, my mind clears.

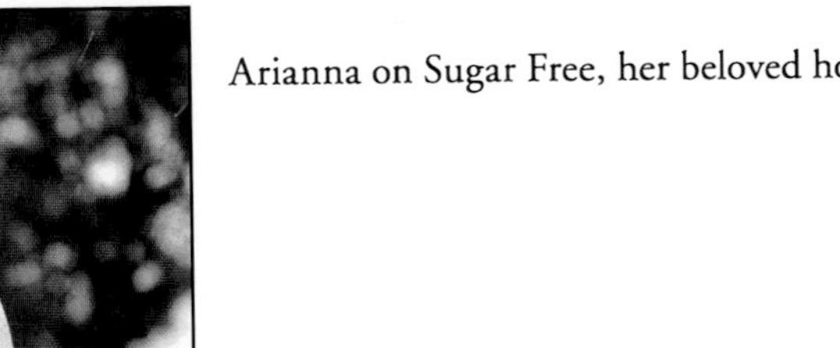

Arianna on Sugar Free, her beloved horse.

Ayla, who was an *American Idol* semifinalist, in concert.

Me, surrounded by the three most beautiful women in the world.

I ran much of my U.S. Senate campaign from the back of my truck.

With Rudy Giuliani during an event in Boston.

A fantastic rally in Worcester, Massachusetts, right before the election.

My first election night with Gail beside me.

Having my family all there was a very special moment.

We all did it!

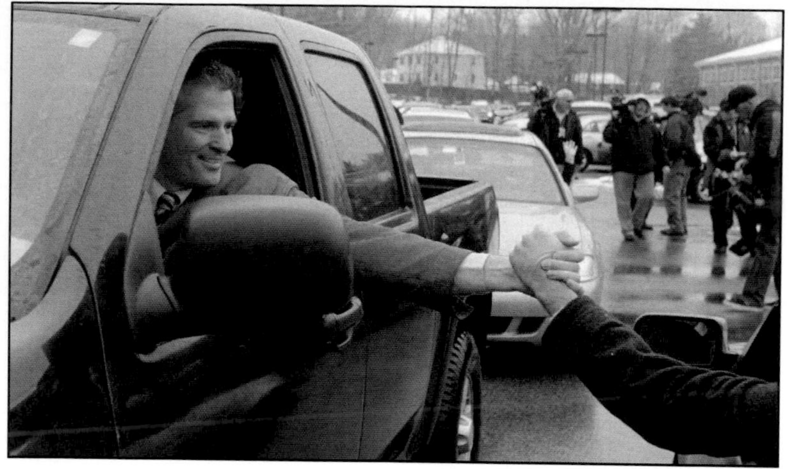

This photo captures my philosophy of public service: every voter, every person matters.

For over a year, cameras and microphones have followed me everywhere.

Thanks to the people of Massachusetts, the truck has come to Washington.

Chapter Ten

ESCAPE

In high school, I didn't do well in German. I didn't do well in Spanish. By my sophomore year, my last resort was Latin, and Mrs. Paula Smith, my teacher, was a major reason why. She knew just how to challenge me, saying, "A big tough guy like you, you can't even get an A on a test. You can't even do well in a simple language like Latin?" And of course, I took the bait. I thought, "All right, I'll show her." I studied, and I worked, and I ended up with A's. Latin, for me, was like the basketball drill of language: everything was fundamental; every word or phrase was possible to break down, like the basics of a jump shot. It was orderly and structured. It required memorization and repetition. And I thrived.

Across the state, Latin and Greek students had formed something called the Junior Classical League. I ran for one of the offices as a sophomore, and as a junior I ran for president. The president would host the state convention—which included mock Olympic games—at his or her high school, and invite participants from forty other schools in the area. It was a real series of games, with a catapult contest and chariot races, but it also included basketball, Frisbee, an egg-and-spoon relay, and academic contests as well. For the opening ceremonies the students, who came from as far as New Bedford and Mount Greylock, were expected to wear togas and to sing the Junior Classical League Song. We also presented a performance of the story of Peter Rabbit, told entirely in Latin.

But my most rewarding moment with the league was a ten-day trip that I took with other Massachusetts kids to the University of Rochester for a convention of high school Latin students from across the country. I would be representing all of the Massachusetts students in the Junior Classical League. Aside from meeting so many kids with similar interests, there were two remarkable things about it, and one was playing basketball in the University of Rochester gym. It was a beautiful, spacious gym with high ceilings and seats to accommodate huge crowds. I was in there one afternoon shooting the ball alone when another kid came in, and I asked him if he wanted to play a little one-on-one. "Sure." We played, and I beat him, as someone watching said, "You kicked his butt all over the court." I had no idea as we sparred over the ball, but the kid I was playing was no kid. He was the captain of the Rochester varsity basketball team, and the team's coach, Coach Neer, was standing off to one side, watching. Afterward, he asked me, "Hey, who are you? Do you go to college here?" And I said, "No, I just finished my junior year in high school." Rochester back then was a Division III team, but it played Division I teams like Ohio State and other powerhouses in the Kodak Classic basketball tournament. Coach Neer looked at me, asked where I went to high school, and said he'd be in touch.

Those ten days, I also met a gorgeous southerner from Alabama, Ann England, who was president of her Latin Club and had been Miss Teen Alabama. For all ten days, we were inseparable, and we were in tears when it came time to leave Rochester and return home. I called her constantly from Massachusetts, and she called me. One month, I think my phone bill was close to $800, earning Larry's immense wrath. It was puppy love, crazy and brief, but I felt it in the deepest reaches of my heart. And it reminded me that there was a life beyond June Circle.

By my senior year I had a motorcycle, which I had purchased used from a neighbor for $300, so I wouldn't have to bicycle everywhere. But I had to be careful where I started it. If I throttled the engine too loudly, it would set Larry off. By now, most everything set Larry off.

On the basketball court, I had become a scoring machine. My junior year, against both Stoneham and Watertown, I shot 27 points in each game. In a double overtime against Lexington, our bitter rival, I scored 15 points in the first quarter, breaking the school scoring record. I now had one more season to prove myself, one more season for the recruiters.

The local papers called me "Duce," playing on my old nickname, Deuce, and wrote about every battle. And they were fierce. Wakefield had been cochamps the previous year. Everyone was looking for us to repeat, but for our second season now, we were in the sights of the other teams. Some teams would do anything to win. They'd play dirty, take cheap shots, try to trip me or someone else and send us sprawling to the ground. They'd trash-talk, and they'd do one-on-one, trying to play mind games to get us off our game. If I went to get a drink, there would also be a kid from the other team there with me getting a drink. If I went to take a shower, the same kid would be right there handing me a towel. Wherever I went, someone from the other team would go. I had people run beside me on the court, two people shadowing me to make sure that I didn't get the ball. They would try to get me in foul trouble, or even to foul out. They'd run alongside me, taunting, "How ya gonna score, Brown?" or "You suck. I'm going to shut you down, Brown." "You're nothing." "You're only going to get five points to-night." But they could never get inside my head. I had too many adults who had already tried to do that.

Instead, I worked harder, and when the ball sank with a swoosh in the net, I'd say, "There's one." And then, swish, "There's another." "There's two more." And then the guys following me would start getting frustrated and would start fouling. They'd foul out, and I'd say, "That was pretty easy. Bring in the next guy, will you?" But what I loved most was when I was up against the other team's best player. Most of us developed a kind of grudging respect for each other. We could score, look over, and smile, and either he or I would compliment the other. Those were the best rivalries, because we matched up one for one and we appreciated each other's game.

That final high school season, we started off beating the Belmont Marauders 60–43, even with their scoring machine, Jay Jehrian. I had 24 points and nine rebounds. Then we were humiliated by Lexington, 84–63. We were fourth in the Middlesex League standings, with Lexington and Winchester and Burlington all ahead of us. My goal was to score 20 or more points in every game, to help us bring home the win.

Against Woburn, I scored 17 points in the second quarter and 34 points in the overall game. I wore my red sneakers and constantly drove to the basket. We beat Lexington the second time in a double overtime, 73–72. Then came Winchester, number two in the league. We were down about 18 points going into the fourth quarter. It was one of our last games of the year, and I did not want us to give up. In the huddle, I kept saying, "Never quit. We cannot quit." And we didn't. I hit 35 points that game, 16 points in the final quarter, breaking the school record for points scored in the fourth quarter. I'm told that the record still stands today.

When the buzzer sounded the score was tied. We lost in overtime, but we had proved that we could come back.

I finished the season having scored 20 or more points in nineteen of our full twenty-three games. My point total for the year was 519, I was only the second player in the school's history to break the 500-point barrier in a single year, and on the court, I averaged 23 points a game. Over three years, I scored 940 points, the second best ever at that point in school history. That season, I was named co-MVP of the Middlesex League and I was invited to be on the Eastern Mass. All-Star Team, in a final high school all-star game that each year pitted Eastern Mass. against neighboring Connecticut's Class A Top 10. The event was played at Southern Connecticut State College in New Haven. National college and professional scouts would be watching us. Connecticut was heavily favored to win. That night, when we stepped onto the floor, there were 2,500 people in the stands.

The Connecticut team took the lead to start, but we drove back. Twice, I stole the ball before halftime, where the score was tied at 33. When the clock stopped, Eastern Mass. had won, 78–69, the first time a Massachusetts squad had ever won the event. Earlier that season, I had been on the court for Coach Lane's one-hundredth career victory. And I had never let Larry break my hands.

The college recruiters came all winter to meet me

and look at my game films. The University of Rochester offered me full financial aid and a scholarship. The University of Maine, Ole Miss, Colby, Brandeis, Connecticut College, and Tufts all came, the low-end Division I schools, high-end Division II, and all the top Division III. Tufts' young coach, John White, had been watching me from the stands since my sophomore year. And when the college team coaches and recruiters came to meet me, I never took them home to Larry's. We met in the living room of Coach Lane's house. He was the one who called them, who sent them the articles about me. He was their contact and the person who steered me through. I loved Rochester. I loved a lot of the schools. But in the back of my mind, I knew that I could never leave the Boston area. If I left, who would protect my mother? Who would protect Leeann? I looked at all the schools, I dreamed, but I knew that Tufts University was in Medford, Massachusetts, its campus running up and down one of those ancient glacial Massachusetts hills. And Wakefield was only a ten-minute car ride away.

I consoled myself at first with the thought that the Tufts gym looked just like the Rochester gym, but then I visited the school. It offered me a significant sum in financial aid, some of it provided by generous scholarship funds from past alumni. The coach was cool, and the team was great. And Coach Lane made one other point:

I was going to be one of the only white kids on the team. Unless it was the all-star game, I had never played on anything but an all-white team. Tufts was a very international school: we had a Korean player, Jimmy Campbell; the Greek players, George Mazereus and John Caragiorgis; an Arab player, Billy Gorra; and Fielo Toro from Puerto Rico. The team was in its own way a kind of mini–United Nations. When I was torn among Brandeis, Colby, and Rochester, he said, "You know what? You know what the difference is between Tufts and these other schools? At these other schools, you'll do great, and you'll love it, and they're great schools. But look around at Tufts, what do you see?" And I said, "I see people." He prodded me. "What kinds of people?" "All kinds of people." He said, "That's right. There're minorities here. You're going to be playing with people you have nothing in common with. When you get older, you're going to be able to relate better to blacks and Puerto Ricans, all types of nationalities, rich people and poor people. And if you go to other schools, I'm afraid you're not going to get that." It was 1977. Not very many people thought that way back then. But Coach Lane did. And I chose Tufts.

But Coach Lane had another reason for steering me toward Tufts. He knew John White. John had grown up as the oldest of seven kids with a single mother in the projects of Somerville, a tough town wedged between Medford and Cambridge, the home of Harvard, across the river from Boston. He had graduated from Tufts, and for him, basketball had also been the only ticket out.

It had not been love at first sight when John White first saw me play. He considered me a skinny little kid, wiry and gangly, who ran slightly pigeon-toed. He thought I had a runner or track competitor's body, not the strength to make me a basketball star. I was going to have to be quicker and stronger, and something other than a jump shooter. But for three seasons, Coach White kept coming back to watch me from the stands. And what he saw was that when there was a big shot to be taken, a big play to be made, a defensive move at a crucial moment, or a loose ball that someone had to dive for, I was in the thick of it. I played better under pressure. I was always willing to take the risk. He later told me, "When night after night, you, Scott, become the physical protector of your mother, when you take those punches, competing on a basketball court is a lot less daunting."

Coach Lane wanted to hand me over to another man who could be a mentor to me on the court and off. He thought the best person for that job was John White.

When the 1977 high school basketball season was over, Coach Lane wrote me a note that began, "Scott,

You've become part of our family." I pasted that in my scrapbook, next to my game articles.

I still saw Brad and Judy Simpson too. Once, before a date with my high school girlfriend, Kathy Donehey, I drove my motorcycle over to their house and hinted that I needed a jacket to wear. Brad produced a tan leather jacket, cut like a blazer, with big round buttons and wide lapels. I was about his size by now, and to me, that jacket was the coolest thing I had ever seen. For two years, even when I was a college freshman, I would come by to borrow it, when it was cold and I wanted to look dressed up for the evening. Brad still has that jacket, its sleeves slightly stiff and cracked, tucked away in a plastic bag in his basement.

By the end of my high school years, my father had ceased making all child support payments. His $25 a week checks, always intermittent, had stopped cold. I arrived on campus at Tufts with nothing, except what I had saved from summer work. I had started painting houses, because I liked being outside, I liked the methodical work of sanding, prepping, and painting—alone with the music on the radio—and it paid well. Or at least well enough so that I had some money for books and a few weeks of food.

I had been heavily recruited for the team, but I was anxious about whether I would play. I had heard lots of stories about guys who came on as recruits and then sat on the bench for the year. I was a freshman; there were juniors and seniors who had been with the team and who were starters. Coach White had been named as head coach of the Tufts team just two years after he himself had graduated from the university. His first year, he was coaching many of his former teammates. Coach was a powerhouse player who was only five feet five and looked like Tony Orlando from the 1970s group Tony Orlando and Dawn. He brought his street-kid toughness with him to the court and made us play for our spots on the squad. His philosophy was that you are only as good as your next game. No one was allowed to rest on his laurels. No matter how hard he recruited anyone, no one was owed anything, and most of all, no one was owed a spot on his team. It had to be earned and won. Every practice was as competitive as a full-fledged game, sometimes more so. A few times, players had fistfights on the floor. That fall, I was one of them. I ended up taking away the position of a good friend of mine from a town near Wakefield, and he didn't like it; we took some swings at each other right on the court. College was also an introduction to the world of adults. Our captain,

Jimmy Campbell, was already married, with a young son.

My mother must have dropped me off at Tufts, but I have no memory of her being there to help me move into my dorm. All of my stuff fit into one small carload. I had a stereo that I had saved hard for, my trophy from being named co-MVP in the Middlesex Basketball League, and some Wakefield banners that I hung on the wall. The room was brown brick, with two metal-and-wood desks and a set of bunk beds that after too many uncomfortable nights my roommate and I eventually took down and set up side-by-side as twins.

My freshman year, I roomed with another player on the team, Benji Williams, a black player from Roxbury, Massachusetts, six feet seven inches tall, who had gone to Don Bosco, a Catholic school. We were polar opposites, but we got along well, even though our lifestyles were completely different. When I was going to sleep, Benji was leaving to go out; when I was getting up, he was rolling in. I had a steady girlfriend back home in Wakefield; he'd bring girls back to make out in our room while I'd be trying to sleep on the bunk below. I was regimented about sleep and studying and working in my work-study jobs at the gym for financial aid and doing late-night cleanup at one of the local pubs. I wanted to get good enough marks so that I could stay on the team and play basketball. Benji sometimes tried to get me to do his work for him, but I couldn't do mine and his. I was practicing three to five hours a day, working, going to class, studying, and then in my free time trying to have a social life. For finals, I reluctantly went home to Larry's dreaded house to study because I just couldn't study enough in my room. But Benji was probably far more typical than me of some college athletes at the time.

The diversity of our team made us unusual in our league, the New England Small College Athletic Conference. In Wakefield, I never had to think about race. The town was about 98 percent white, but our team at Tufts had a fair number of black players, and suddenly, I began to see things through their eyes. We played most of the top New England schools—Colby, Bates, Williams, and Amherst—as well as a host of smaller ones. Most of the teams weren't integrated, and when they did have African-American players, those guys usually were prep school grads. Our players were like Coach White, from the inner city, and they had honed their skills on the street. When we traveled to other schools, particularly the more remote and insular ones, people in the stadiums sometimes spat on us or called out racial slurs. A lot of it was horrible, vile, vicious language, and each time, I would think, "Damn, I'm glad I didn't go here."

Many times, though, the prejudice was far subtler. When we played the Bentley College Falcons in 1979, their game program had a little write-up on their evening match against Tufts. The featured photo was of our three African-American starters, with the caption "Roxbury Connection," adding, "The three Roxbury residents are the key to Tufts' inside game." Roxbury, of course, was the "black section" of Boston. It had been the scene of riots and looting following the assassination of Dr. Martin Luther King, and by the late 1970s, its streets were a mass of vacant, trash-strewn lots and the burned-out shells of old industrial buildings, many gutted by arson. In a few years, it would become the epicenter of the city's crack epidemic. To highlight our team's "Roxbury Connection" was to offer multiple meanings. In our league, other teams did not have inner-city players. We were the exception to the rule. But the stream of comments also had another effect: they made us tighter as a team. We didn't have cliques of people; all the guys hung out together. People who saw us play either loved us or hated us. There was no middle ground. We knew that—and nothing ever pushed us apart as a team. Ever.

And we came to notice one other thing. When our

team traveled to these other schools, in the stands, their few African-American students were often rooting for us.

I held my own that first year as the team's "Sixth Man," the guy right behind the five starters. I was able to become one of the top four scorers on the team—when I scored 22 points in an overtime loss against Sacred Heart, the Eastern College Athletic Conference named me Rookie of the Week. The next week, we were slated to play MIT. I started, but after four minutes Coach White pulled me out. That was it. I didn't play for the rest of the game. Coach didn't tell me why; he didn't say anything. After the game, I walked up the stairs to the locker room. I showered and changed, and when I walked out, Coach White was standing there, being interviewed postgame. I looked at him, dropped my uniform on the ground, and walked away. All those hours in the gym, working, studying, practicing, they all collapsed in that uniform on the ground. It was the first time I had ever walked away from anything. I walked through the gym, down to an alleyway off the indoor track underneath. And I began to cry. I stood in that alleyway, under the stairwell, bawling. The only thing I wanted to do was play; inside, I just felt hollow, as if everything had been drained from me. I was still crying when Coach White found me.

I told him that I didn't know why I didn't play. I was ready to win, ready to contribute, I had been playing great. And then he didn't play me. I played for four minutes. And he told me that he didn't want me to get a big head, to get too full of myself. I looked at him, stunned. "What do you mean, 'full of myself'?" I asked him. "I'm working my ass off trying to be the best possible player I can be." I said, "Do you know what I go through? I go home. I take care of my family. I'm studying. I'm working. I'm never late to practice. I always stay long after practice is over to keep working on my game. And you're afraid I'm going to get full of myself?"

He handed me my uniform and said, "Well, why don't you pick up your uniform and think about it?" I did think about it, and I didn't quit. As with everything else, I showed up at practice the next day and worked even harder. I scored and contributed more. Years later, I saw the lesson, to accept adversity in whatever form with balance and grace. I never cried again over sports. I took the challenge to allow that game to make me better, and to make me a better master of myself. With even more hours of practice, I could make everything go right on the basketball court.

Where I couldn't make it go right was at home.

Leeann was usually the one who called, at night, and she would find me at the gym, or in my room. She would be crouching in Larry's front hall study, up on the second floor, in tears, whispering into the phone. Larry had started in on mom again. Sometimes, she would be locked in the bathroom. But sometimes, when Leeann was on the phone, Larry would have his hands around my mother's throat. With me gone, there was no check on his rage, and no other outlet. He could pin her to the wall and do what he wanted. I would drop the phone and race for my motorcycle or a car, borrowing the keys from my coach or someone on the team. Leeann would be screaming to Larry, "Scott is coming!" That was usually enough. If I arrived when he was still there, he and I would go at it: he would come at me with his sawed-off fingers; I would throw him up against the wall in the hallway. But many nights he just took off, and I would race home to find him already gone. The only visible residue of his rage were the phones constantly ripped out of the wall sockets, their dangling cords and shredded wires streaming helplessly on the ground.

I might go out looking for him, riding around Wakefield, searching in his familiar haunts, or I might sit in the living room and wait for him to return, for the usual round of apologies—his "I'm sorry," followed by some excuse. He always had a smooth line afterward, something almost hypnotic that drew my mother, Leeann, and me back in. And truly, where did my mother have to go? She had no money, no profession, no child support from Leeann's father. Larry controlled every bit of cash in the house; he paid the expenses for Leeann. A divorce lawyer would require a retainer of thousands of dollars, and my mother had nothing to her name. She had very little sense of any of her rights, and Massachusetts law back then gave her very few options for where to turn. Larry ruled their lives, to the point where both my mother and Leeann were sometimes too scared to tell me what was going on when I dropped by just to check on things. But I could read it in their faces, and I could read it in the smugness that radiated from Larry's eyes.

After each episode, I left, knowing that I would be back in a few weeks or a month or two. The promise that things would be different was nothing more than a fragile truce that would hold only until the next time. There was no way to curb the violence. It was gathering out there like a storm, and every night I waited for the phone call, never knowing for sure when it might come.

By my sophomore year, Larry was routinely physically and mentally abusing my mom and sometimes my sister. On occasion, he would lift my mom against the wall and lean forward into her face, slowly crushing her windpipe under his forearm or his palm, choking her. And he would hurl Leeann against that same wall. I remember racing home during one of these incidents and finding him in the hallway. This time I didn't care. I was not afraid of him. I was big now, 185 pounds of solid muscle from lifting with the Nautilus machines. I pinned him against the wall, the bone of my forearm against his chest, my left fist pulled back and balled, ready to connect, and I told him, "You fucking touch my mother again, you fucking touch my sister again, and I will kill you. I will absolutely kill you." And I was ready to. I didn't care what happened. I knew that the next time, it would be my mom or Leeann. For a while, he got the message.

I was lucky to be on the basketball team because that gave me free clothes. I got shorts, T-shirts, and sweats. And that was what I wore, to class, to parties. I had a jacket, a hat, and a sweater, which I had to make last a very long time. I would do almost anything for extra cash. In the dorms, the resident advisers or even Coach White, who lived with us with his family as a faculty resident, would call me and say, "Hey, Brownie, someone hurled on the stairs. Can you come clean it up for ten bucks?" And I'd be right there, with a shovel, a bag, some rags, and disinfectant. Two minutes of cleanup for some drunken vomit, and I'd have ten bucks. I could get through the week on ten bucks, and a lot of weeks, I had to. I would do anything to make money; no job was too disgusting or too small.

My mother couldn't help with much, and my grandparents were in New Hampshire, but Gram would chip in with what she called pin money, sending me an envelope with $10 to $20 every once in a while. I was working, practicing, and studying all the time. I was on full financial aid, but I had no extra money to live on or to go out with friends. According to the court papers, my dad was obligated to continue his child support payments through college, if I attended. But even when I was in high school, the $25 a week checks had been intermittent. Now, they were entirely gone. I don't know where I got the idea to sue my dad to enforce the agreement, but I did. I went to the local courthouse and looked through the legal documents. On my own, I collected the right papers and I prepared to file them, not only asking for the $25 but petitioning for $50 a week because of "a change in circumstances." That change was college. I called my father and told him, "Dad, I need some money. I have no money. I need money just to eat. I need you to continue the $25 a week, $100 a month, or even to up it to $50." His answer was, "Are you kidding me?" He wanted to be done. It was like pulling teeth to get him to honor the agreement. Finally, he grudgingly relented, but the payments stopped again after about a year. The second time, I gave up. Even though he was back in Newburyport, opening a country store with needlepoint pillows, scented candles, and painted trays with Linda, his third wife, my father didn't come to many of my basketball games. The relationship was too strained. Over $25 a week.

Instead, Coach White and I went back to the financial aid office and begged for more money for me, so that I could cover books and a meal plan. But as bad as the coach knew things were, there was a lot more that I never told him.

The worst night of my life came toward the end of my sophomore year. That night, once again, Larry had my mom up against the wall. He had his hands around her throat, but this night, he pushed harder, until her face was turning blue. Leeann saw her eyes begin to roll back in her head and she, age thirteen, rushed Larry from behind. She kicked him in the balls and began hitting and biting him. It was like me and Dan Sullivan all over again. Larry whirled around and threw Leeann up against the wall, with a force so violent and strong that her head cracked against the Sheetrock and she almost lost consciousness. But to do that, he had to let go of my mom. My mom broke free and raced to the phone in Larry's study, locked the door, and called the cops. She also called me. The cops came and escorted Larry out of the house, and I came back home, to stay for a while. Either by court order or by some other arrangement, Larry was forced to move out. But he was not totally gone.

He was still allowed to come around to see his mother, who continued to live in the in-law apartment above the garage. He was also responsible for all the utilities and the maintenance. But that gave him another form of control. In the winter, he would periodically make sure that the oil was not delivered, so that the heat could not be turned on in the house. He would let the electricity bill lapse until the power was shut off. He did the same with the phone. When he lost the proximity to physically abuse my mom, he tried every type of mental torture on my mother and Leeann. One night I got another hysterical call from Leeann. Her dog, Taurus, was missing. Larry had taken him.

I got into a car and began looking for Larry. Finally, I found him in downtown Wakefield, in the second floor of a building he owned, with a thin, tenement stair stretching up the back from the parking lot. He had the dog, an animal that he would periodically hit with his scarred palm when he had been at home. I went and got a friend of mine who was a cop, and I asked him to come with me to get the dog back, because I was afraid of what might happen if I had to confront Larry, just myself, one on one. We retrieved the dog. Leeann and my mother stayed in June Circle because my mother had nowhere else to go. When they finally did leave, going to an apartment, my mom was back where she had been nearly a decade before, a single mother on welfare, struggling to make ends meet.

Once I knew that Larry was gone for good, I stopped coming around as much. I stayed in school as long as possible, until the very end of exams, leaving late for breaks and coming back early. There was only a small space for me in my mom's new place, just enough room to crash, but I also needed the distance. I was tired, spent, worn-out down to the bone. I had lived my entire life under the shadow of other men and other choices, choices made before I was even born.

My sophomore year two other things happened: I scored 35 points in one game over Bowdoin, and I fell in love. Her name was Pam; she was from Lexington,

Massachusetts; and she was a tall, dark-haired tennis player. Her parents were divorced, and she was from a very athletic family. Sports and familial chaos, albeit on a far more manageable scale. We were almost inseparable. I would hang out around the tennis courts in the center of the campus and watch her play; she would come to my games and cheer. We shared meals, gym time, and runs, and I could talk to her just enough about the chaos with Larry. At the start of our junior year we moved into a house off campus with about six other people, in one of the blocks of narrow, boxy rentals that owed their existence to the steady influx of Tufts students—with marred walls and used furniture, and the smell of old food and stale beer. We thought that we were mature and independent, and we had long, serious conversations about getting engaged and married. I could see my life mapped out, down to kids and a dog waiting for me by our front door. We spent a lot of our free time at her Armenian grandma's, eating traditional cooking. But Pam's mother especially didn't want us to settle down. She was forever trying to discourage me and to encourage Pam to try new things. One of those things was six months in England. We wrote, and I flew over to visit, my first trip ever that did not involve basketball or Junior Classical League, but from the moment I arrived, I could feel that the ground had shifted. It looked as though she was dating other people. For two twenty-year-olds, an ocean provided too much time and distance. Sadly, our relationship ended, albeit amicably, soon after she returned.

I was a good shooter on the team, with a chance the next year to be voted in as captain. The school paper called me "Downtown Scotty Brown" because I liked to shoot so far away from the basket. And I had found my path. I was making the dean's list for my grades, and I had run for the student senate. My platform was halting tuition increases, being more responsible about how student fees, particularly social fees, were spent, and improving dormitory safety, including fire and theft precautions. "I won't promise to change everything," I wrote for my statement, "but I will promise to be a sincere, active representative of our student body. Please consider me!!!" I was number forty-five on the ballot, out of fifty-seven candidates, and I won.

But the most important thing I did wasn't in school. It was on December 29, 1979, in the middle of my junior year, when Pam was gone. That day, I went to the armory at Camp Curtis Guild to officially register to join the Army National Guard.

Chapter Eleven

THE *COSMO* GUY

In February 1978, while I was a college freshman, a huge nor'easter rumbled up the coast of New England. The storm began as a tropical cyclone off the coast of South Carolina, and as it rotated north, an Arctic cold front merged with the rain. The water turned to snow, but it never lost its tropical orientation. It arrived in New England as a blizzard with hurricane-force winds. Across Massachusetts and Rhode Island, the snow fell for thirty-three hours, coming down so fast and furiously that cars were literally trapped on the roads, blocked by blowing drifts. Some 3,500 vehicles were abandoned in the middle of streets and highways. People became trapped in their cars, some dying because the snow's rapid accumulation blocked the tailpipes and the carbon monoxide fumes killed them. Thousands of homes and businesses lost electricity, heat, and water. Snowdrifts piled up as high as fifteen feet. All road traffic was banned and Governor Michael Dukakis called out the National Guard to help clear the highways so snowplows could pass.

The Guard was my first experience with the military, aside from my neighbors in Malden who had served in Vietnam. I watched the images of the Guard members in the bitter cold, rescuing the stranded, clearing the roads, and I was really impressed. So much so that the following Thanksgiving when I saw one of the local Guard commanders at the Wakefield–Melrose football game, it was all that I wanted to talk about. He told me I should think about joining and began trying to recruit me. At the time, I was largely thinking about how my mother and Leeann could get out of the house on June Circle, away from Larry's reach. But I kept the idea in the back of my mind. By the fall of 1979, I was ready. Fifty-two Americans had been captured and were being held as hostages in Iran. Then, on December 25, 1979, troops from the Soviet Union invaded Afghanistan. For me, that sealed the deal.

I was determined to graduate from college and to play my final basketball season, but I wanted to join the military. I wanted to serve. Here were men whom I could admire, men whose job it was to protect others as the normal course of their lives. There were so many undercurrents behind my decision, many of which I never would have recognized then. But standing in that armory, I felt I belonged. I signed my papers, knowing that if I were called to active duty, I would go. And if I stuck it out with the Guard, I had the added bonus of knowing that it paid a salary to those who participated. From my days scooping vomit, from my every summer painting houses or mowing lawns or doing whatever odd jobs I could scrounge, I knew that a couple of thousand dollars might be the difference between going on to some kind of graduate school, probably law school, or not going.

At Tufts, Coach White was shocked. With the Guard, I was expected to spend my summers training. How could I play basketball? How would I practice over the long summer months? I hadn't really thought of that; I'd just assumed that I could play, and the Guard hadn't dissuaded me. My parents, who never weighed in on anything, each independently asked me, "What the hell are you doing?" But my mind was made up.

I was supposed to report for two months of basic training in June, at Fort Dix in New Jersey, where I would be joining Bravo Company, the Third Battalion (BRAVO 1-3). Fort Dix was where thousands of U.S. troops had trained before being deployed to Vietnam. The base had even built a mock Vietnamese village on its grounds. I wanted to be a top-notch soldier, and I decided to train for basic training. To prepare, every day I ran the three-mile loop around the lake in Wakefield with a full backpack or a duffel bag and I did a ton of push-ups, sit-ups, and pull-ups, so that on day one, I would be ready.

We arrived at Fort Dix just at the start of the Guard's efforts to fully integrate female soldiers, so on our part of the base, in my barracks, the top floor was for women and the bottom two were for men. The drill sergeants came down hard on us, threatening that if we got any of the women pregnant or were caught hanging around with them, we would be out of the military. There was no fraternization at all, although over the summer, a few of the drill sergeants did try to move in on some of the more attractive female recruits.

I quickly became a favorite target for the drill sergeants because I was a college guy entering basic training as a private first class, and not the typical Guard recruit. It was a little bit like my years with Coach Lane—the sergeants could yell at me first. On one of my very first mornings, I was standing in the front row when a six-foot-four drill sergeant walked up to me. He was African-American and sewn across the front of his uniform was his name, "Brown." I gave him a little smile. I didn't say anything—I just turned up the edges of my mouth quickly, in a kind of "Hi, how ya doing?" expression. But the military is not for smiley guys. He immediately called over to the other drill sergeant, "Jonesy, come here." The two of them started circling me like a couple of sharks. "Jonesy, do you notice anything about this young soldier?" "Yeah, I do," came the reply. "His last name is Brown." "That's right," said the drill sergeant. "What's my name, Jonesy?" "Brown." At that moment, Sergeant Brown looked me right in the eye and said, "You know, I think we're related, PFC Brown. No, PFC Brown, I know we're related. We have a distant relation in our family and as a result Mom told me to look out for you. So I'm going to do just that. I'm going to be all over your shit every single day." And I said, "OK, drill sergeant."

He paused. "Hey, Brown, you know what? Why don't you get down and give me ten?" So I dropped down and banged out ten quick push-ups. And I said, "Permission to recover, drill sergeant?" He said, "No, no, no. Why don't you get down and give me fifty?" And I dropped down and banged those out too. That caught Sergeant Brown's eye. The next time the drill sergeants told everyone else in the group to give them ten, I would have to drop down and give fifty or one hundred. From that moment on, when it came to me, Sergeant Brown was always watching.

Basic training was a daily competition in the humid New Jersey heat. We awoke at 4 a.m. to do push-ups, sit-ups, running, and drilling. We ran in our combat boots—I did a five-minute mile in them. Back then, it didn't matter if the boot's soles were hard or if the seams cut blisters or if the skin on the bottom of our feet peeled off in long white strips. We ran in those boots every day. Five thirty a.m. until six was the time to do what was called "shit, shower, and shave." At 6 a.m., we were in the mess hall, shoveling food into our mouths. If you ate too slowly, the sergeants came over and made you dump your plate. Then we were off to training.

During our exercises, we learned how to fire rifles (like the M-16), shoot pistols, and hurl grenades. We learned how to bivouac, set up tents, and navigate in the woods. We crawled in the mud beneath ropes as live fire whistled over us. We jumped logs and rappelled down walls—everything you see in the movies—except that

I and all the other recruits were in the front row, caked with dirt and dust, mosquitoes biting every inch of our exposed skin. And the weapons in our hands were real.

Because I had over two years of college when I joined, I was inducted into the Guard as a private first class, and when I arrived at Fort Dix, I was put in charge of a platoon, about eighty people. Most of the recruits were two or three years younger than me; they were primarily African-American, Hispanic, and southern kids. There were a bunch of kids in my group from the Appalachian mountains, kids who came from West Virginia coal and Virginia and Kentucky tobacco country, who grew up in hills and canyons and for whom New Jersey was a very long way from home. I had never set foot in the South, not even Florida, and I could barely understand a word they spoke, but no one could outshoot them with a gun. Especially the guys from Alabama. They could hit any target—the easy ones probably with their eyes closed.

And like me with my basketball, living in Larry's house on June Circle, some of them were looking for any way out of the place that they called home.

Leading these men was one of the hardest things I have ever done in my life. I was the guy they came to with their personal problems, who made sure when they went on leave that they came back and didn't get into any trouble, who spent part of his evening getting everything organized for the next morning at 4 a.m. I had to lead. Leadership required not falling prey to certain facets of military culture in 1980. Chief among them was what many called "buddy-fucking," a phrase that has almost no equivalent in civilian life. In the do-or-die culture of the Army, it meant people who are nice to your face but who are trying to screw you behind your back. Any officer of any rank, even a private first class like me, heard it and worried about it almost from the moment he or she stepped on base at Fort Dix.

If you were a platoon leader, you got your own quarters, as opposed to having to bunk in a group with the rest of the men, and all the platoon leaders, including me, slept with footlockers or chairs in front of their doors. At night, some of the recruits who wanted to cause trouble would have what they called blanket parties. They'd come in, pull a blanket over your head so that you couldn't see, and pummel you. The first time I went out on leave, I came back to find my room trashed, my stuff ripped up and soiled. I cleaned it up without a word. After that, I hid my stuff whenever I went away. It was part of the nature of the beast, a military still demoralized after Vietnam, still trying to find its way.

I, in turn, also pushed the envelope. I tormented my drill sergeants because I struck up a flirtatious friendship with one of the female trainees, a southern girl with brown hair and crystal blue eyes. The sergeants made it their personal challenge to catch me, and I made it my personal quest not to get caught. We'd sing together in church on Sundays just for a chance to hold hands, and the sergeants would threaten me with KP duty or say that they were going to take away my leave. We'd wink at each other, we pushed every rule to the edge, but we never got caught. Then came the sit-up contest.

One part of our daily training was physical training, PT, which included sit-ups. Each morning, we would drop to the ground in a group, men and women, and be expected to crank them out. We started with one hundred, then went to two hundred, then three hundred. Other trainees dropped out, but I kept going, as did my quasi-girlfriend with the crystal blue eyes. Soon we were up to five hundred sit-ups, then six. Now the drill sergeants were interested. They were clustered around in their combat boots and fatigues and brown T-shirts stained with sweat. And they were egging us on—who is better, Brown or the girl? Who's better? If they could have peeled off a wad of cash and made wagers, I'm sure they would have. I kept going, and she kept going. We laid our backs against the hard ground and then pulled up our torsos, contracting our stomach muscles one by one. The ground began to scrape against my spine and chafe against my tailbone. My eyes stung from the perspiration dripping off my forehead, but I kept my hands locked behind my head. I didn't even try to wipe it off. We got to 1,500. Neither of us would quit, and a crowd was watching. Sometime after we had crossed the 2,000 mark, she gave up. I managed five more and then collapsed, my tailbone raw and bleeding, my entire backside and stomach writhing in pain. And even after that, we kept flirting and pushing the envelope. It was my outlet for the summer. That, and basketball.

In the summer of 1980, Fort Dix was still a training base for regular U.S. troops, and during my free time, I would sneak away from the basic training camp to go play basketball with the men in the Regular Army. I kept my basketball sneakers hidden away in the bottom of my laundry bag and my shorts and T-shirt on underneath my uniform. While the other recruits were out drinking, I was hustling up basketball games, spending a couple of hours on the court.

After my two months were over, I had earned some awards—including a distinguished athlete award and the "Trainee of the Cycle" citation, given to me out of five hundred soldiers—but the biggest one I got, I could never pin on my uniform. When the summer was over, Sergeant Brown came up to me and said, "Anytime you want to go to war, son, I'm with you. You're a college guy, you won awards. You're squared away. If you became an officer, I'd follow you anywhere. You have the makings of a great soldier."

Sergeant Brown got me thinking about training to be an officer. When I returned to Tufts, I looked into doing ROTC, and I began in the simultaneous membership program (SMP). I was in both the Guard and ROTC, doing both at the same time. When I joined ROTC, I got promoted to sergeant E-5. And my training would begin in earnest after my graduation.

Not long after the start of college, it became clear that I had stopped growing. I was six foot one, but I was not going to tower over anyone. Most of my opponents could look down on me. Even my father was taller. There were not many six-foot shooting guards in the NBA from the New England Small College Athletic Conference. I might have been able to play for a season or two in the European leagues, but a professional basketball career was nothing but a pipe dream. I was in my third season as a starter on the Tufts team, and I had learned to play smart, to beat my opponents by thinking two and three moves ahead of them with the ball. If they had height, I went for speed, for consistency, and for muscle memory as I spun the ball toward the net from the corners of the court. I began my senior year with a shooting slump, but by the time my last season ended, I could boast of three college games when I shot 35 points overall in each individual game. The last of them was against Brandeis: I brought home the win with a final-second shot at the buzzer, from thirty feet away. I was closing in on having scored one thousand points over the course of my Tufts career; my final figure was 965. I had been picked as the team's cocaptain, and I was considered one of the best outside shooters in New England's Division III. In the locker room, I even persuaded my teammates, who had very different taste in music, to listen to Queen. Sometimes, to warm up, we would blast "We Will Rock You."

But I had a life off the court as well. I joined the jazz choir at Tufts, where rehearsals were at 7 a.m. I joined in part because one of my buddies, Rich Edlin, was a member of the choir and he convinced me that the six girls in it were all gorgeous. We performed in clubs around Boston, we would go out singing and drinking, and a few times afterward, we stood outside the dorms of some of our female friends and serenaded them. As word of our impromptu midnight serenades spread, we began to get requests. Friends and even strangers would leave notes or messages on our answering machines and ask us to sing outside Wren Hall or some other spot at 12:15 a.m. By the time I graduated, these singing events had become a tradition.

The singing was an outlet for the tension of the court and for the inevitable question of what came next. As I had done in high school when I played the role of Patrick in *Mame*, I appeared in a college production of *A Funny Thing Happened on the Way to the Forum*, playing one of the leads, Hero. My intellectual life blossomed. The guy who once worked to get a B-minus in Political Parties now had a 3.5 average. My favorite class was Yiddish Literature with Professor Sol Gittleman, where we read the classics of Isaac Bashevis Singer and Sholem Aleichem. I had joined a fraternity, Zeta Psi, which had a combination of jocks, intellectual kids, and theater kids. Some of the fraternities were all sports, like football or sailing, but Zeta Psi was a mix, without being cliquey. We had guys in ROTC, guys who were going to be engineers or lawyers. And I, the guy who once stole a suit to wear to a dance, was the Rush chairman.

In the Guard, during the school year, I became a legal clerk, keeping the minutes of any courts-martial or any other legal proceedings, and I was picked to be a deputy probation officer for young offenders in Somerville. I knew I wanted to go to law school, and I was training myself as best I could. But most of all, I knew that but for my coaches and Judge Zoll, I could have easily been one of those young offenders. There was nothing preordained about my path at all.

By my senior year, my mother and Leeann had moved out of Larry's house. It had been a battle of courts and lawyers, with Larry leaving and then being allowed to return, and then my mother leaving. In those days, there were few laws to protect abused wives and single mothers. Now my mom and Leeann were living in an apartment in neighboring Melrose, and Leeann was enrolled at Melrose High. Larry stayed in Wakefield for years. He had his properties, and I would see him sometimes around town, where I gave him a stare and told him to get the hell away from me. He was, I know now, a marital terrorist, as bent on destruction as the guys who build IEDs or wire the detonator on a suicide vest. He used everything at his disposal: violence, money, and physical and mental coercion. He nearly killed my mom, he wrecked Leeann's life for years, and the worst that happened to him was having to pay a divorce settlement. Otherwise, he kept everything he owned. His bonds were safe and he continued to walk the streets with his terrifying partly amputated hands.

I met Ruth Greenfield on a bitterly cold late-fall night at Houlihan's, a bar in Faneuil Hall in Boston. I had arrived with a bunch of my basketball teammates, and I was wearing a scarf—a really long wool scarf, the kind that could be wrapped repeatedly around my neck—because it was freezing. In a 1980 way, I thought it looked really cool. Ruth was already in the bar with a group of her friends from Boston University. I don't remember anything about her friends, just that she had beautiful, long, wavy strawberry blond hair. I said something to her, she laughed, and we spent a lot of the evening talking. She gave me her number, probably on a matchbook, and I called.

Ruth was from Long Island, New York, and she was studying to be a psychologist. Her father had died when she was younger; her mom had remarried, and Ruth and her entire family made me feel special, made me feel wanted. Even her interest in psychology was perfect because I came with my own set of baggage. Ruth was funny, and she wasn't scared off by the fallout from Larry or the other ref-use of my parents' periodically chaotic lives. We dated for the rest of my senior year; she came to my games and cheered in the stands. My parents came too. My mother came to most of my home games, my father to a scattered few by the end.

But the court was bigger, and they weren't on pullout wooden risers at the side of the gym. I couldn't scan for them in my peripheral vision; all I could do was concentrate on the ball. I wanted to be the guy taking the last shot of the game, putting the ball in flight just before the buzzer.

My parents were both there for my graduation. I was one of thirteen seniors to win a "Senior Award" from the Tufts Alumni Association at graduation, and I also earned a one-year, $2,000 NCAA scholarship, which helped to pay for my first semester's tuition at Boston College Law School and got me a spot in something called "Tufts University's Cavalcade of Champions." In my scrapbook, I pasted a Xerox of a photo of me at graduation, in my white suit jacket with the spreading lapels, the round collar and wide tie, a boutonniere pinned to my buttonhole. I'm holding my diploma, and next to me on one side is my father, beaming, the perfect touch of silver at his temples. My mother's eyes are rotated as far away from my father as possible; her face is forward, but she is looking off to one side. On her face there is a hint of a smile. In the middle, I have that frozen expression of someone waiting for the shutter to snap and the moment to be over, and for us to separate and move on.

I had joined ROTC and spent part of my summer training and catching up on my classes, as well as painting houses and doing whatever I could to make some extra money. Boston College Law School started in the fall. I took my first-year courses at BC and first thing in the mornings headed over to Northeastern University for ROTC courses. My fellow cadets were nearly all undergrads, and I was about four years older, gramps to most of them. I got up at 5:30 a.m. and went to Northeastern, several miles away, to train from 6 to 9 a.m., then headed back to law school, and studied at night. And I was dating Ruth. For the first time, my life was balanced. Every piece of the puzzle fit. It was like that perfect stillness, that calm, bright blue sky that surfaces after a raging thunderstorm. And then there came a magazine.

Cosmopolitan magazine, or *Cosmo*, was searching for "America's Sexiest Man." There had only been four previous winners, and they included Burt Reynolds, James Brown, and Arnold Schwarzenegger. I didn't read *Cosmo*, but Ruth did and so did Leeann. The winner would get $1,000, which would partially replace my NCAA scholarship for the upcoming year to help pay for law school. Ruth and I talked about the contest, but Leeann took it further and thought that I should enter. She sent in two photos of me—one in a shirt and a tie and another in a bathing suit. She enclosed a nice letter, telling the judges how I was a law student and in the military.

Months later, it was reading period before finals at BC Law School. I was studying, and my phone rang. The woman on the other end said that she was Helen Gurley Brown. I had no clue who Helen Gurley Brown was. I was sure it was a prank. So sure that I said, "Yeah, right." And hung up. A minute later, my phone rang again. I was still incredulous. I said, "If this is for real, send me the ticket to New York." The next day, a FedEx truck pulled up. Inside was an envelope with a ticket. I was headed to New York.

What hit me first was the size and the overwhelming noise. Boston in 1982 was a small place with relatively few towering buildings. The wharf area had yet to be redeveloped. The city was old and small. New York was neither of those things. It was a cacophony of high-rise construction cranes flocking up the East Side and tall buildings that blocked nearly any patch of the sky. I walked down Madison Avenue looking up, feeling like a mountaineer in a canyon with giant stone formations rising on either side. There was the deep-throated rumble of buses belching thick, acrid smoke and yellow taxis snaking in and out of lanes, their drivers leaning on their horns. The sidewalks were crowded with people jostling by. I met with Helen Gurley Brown and a bunch of other people at the magazine. She looked at me and said, "You're our winner." I said, "You haven't even seen me without a shirt or clothes," and she replied, "It doesn't matter. I can tell you are not going to embarrass us. You're in law school, in the military, and you're a nice young man."

Soon afterward, they set up a photo shoot for me. I was just coming off exams. I was in pizza, beer, and popcorn shape—not fat, but not cut either. I wasn't playing competitive basketball every day. Suddenly, I was on location with stylists and makeup people touching me and arranging me, spraying my skin and my hair, and a photographer telling me to drop my chin and then lift my chin as he clicked away. Previously, the most formal of all the pictures I'd ever posed for involved me balancing a basketball on my index finger while I wore my basketball uniform. Here, I was completely nude, although my privates were hidden, in a roomful of strangers, men and women. I was totally unprepared. I was embarrassed and very, very uncomfortable, discreetly trying to cover myself, and feeling completely freaked-out. It showed. The photos were not what they had wanted. On top of that, I looked pale and I wasn't physically toned; I probably needed to lose ten pounds. The *Cosmo* staff saw the photos and told me to come back in two weeks. They wanted me to be, as they put it, "more cut." They'd be doing a reshoot, at a house in the Hamptons.

I'd never had to get in shape like this before, and I reverted to what I'd done to prepare for the military. I ran every day with a backpack, and I stopped eating. My food was tuna in a can, three times a day. I went to a tanning salon as well and lay under the UV lights. When I returned to New York, I was bronzed and toned. *Cosmo* had its photo shoot at the beach and planned a red-carpet rollout. I was to start in New York and then go on a thirty-two-state tour. *Cosmo* put me up at a fancy hotel and I was scheduled to appear on wall-to-wall media, including the *Today* show with Bryant Gumbel and Jane Pauley, *The Phil Donahue Show*, the network evening and nighttime magazine shows, and even something with Barbara Walters. It was the first time in *Cosmo*'s history that a regular guy had won their contest, and in part it was because, as Helen Gurley Brown had said, they thought that as a law student and someone in the military, I wouldn't do anything to humiliate them, which was true. In the write-up that went with the photo, they quoted me as saying that I'm patriotic, even though it isn't considered cool. And they also asked me about my dad, and I told the truth, that "he wasn't around much." The article also said that I liked "slinky women." It was my first experience in learning not to believe everything you read. Not only didn't I say that; I didn't even know what a slinky woman was.

The rollout was a whirlwind of being whisked from one spot to another, hopping into and out of town cars that raced around the city from 5 a.m. to 11 p.m. I sat in the studio with Bryant Gumbel, and he rattled off his list of questions: What kind of reactions had I received from my friends, and was I embarrassed by all the attention that the photo shoot had generated? Offscreen, in the greenroom, I remember being asked one more question: "If you run for political office, do you think this is something that will be used against you?" I answered, "No. And I don't have any interest in politics."

Outside New York, the national tour was even crazier. Strange women would knock on my hotel door late at night. In Detroit, after one show, a married woman invited me to her home to hang out with her and her newly single friend. There were giant posters printed of me and in a certain sphere of the young, hip New York City world, circa 1982, I was something of a known name. But I wasn't Scott Brown; I was "the *Cosmo* guy." At

Studio 54, the Underground, the Red Parrot, and Plato's Retreat, no matter how long the line straining against the ropes—lines of bankers in their flashy suits and glinting shoes, ties off, shirts unbuttoned, or girls in their shiny satin tanks and skintight Calvins, so molded to their legs that they couldn't sit down, teetering on sky-high Candies—I could walk right through and be ushered in. "Hey," the bouncers would say, "it's the *Cosmo* guy." The first time I went into Studio 54, club owner Steve Rubell and Calvin Klein tried to rip my shirt off as kind of a prank to "see what you got," as I went in the door. I was pissed; I didn't know who they were or why they were doing it. And I didn't own many shirts.

Another night, early on, I was ushered straight through the main club, with its pulsating music, vibrating lights, and people dancing or wandering with drinks permanently attached to their hands, to the back room, where the 1970s disco star Rick James and a couple of friends were sitting in a half circle. On the tables around them were piles of cocaine and draped over their chairs were leggy women with plunging necklines. Someone must have told them who I was, because they called out, pointing to the drugs and the women, "Hey, *Cosmo* guy. Want some of this? Want some of that?" I shook my head: "No, I'm good." And I ordered an orange juice.

That was the benefit of growing up around men like Larry and Dan Sullivan, of my mother's episodic binges, the times she was passed out, the way a night of drinking set off a downward spiral in her or in one of her husbands. I never drank hard liquor. I couldn't stand the smell, and the taste to me was personally revolting. I stuck to beer and never more than two or three. I always wanted to know exactly where I was, to know exactly what was going on. The drugs too I had no interest in. And they were everywhere back then, not just in the back rooms of New York clubs. In a year or two, college sports teams up and down the East Coast would celebrate the end of the season with lines of cocaine. Len Bias, the second overall NBA draft pick, who signed with the Boston Celtics, died of a cocaine overdose before he ever set foot on a professional court. After that first summer in New York, a marketing agency set me up, as the *Cosmo* guy, with an all-American looking, famous hair-color model, who liked to snort heroin. That fix-up didn't last even half the night.

I went to the clubs and the cool places because they were new and I could get in and everything was about being seen. I had been persuaded to give modeling a try.

From *Cosmo*, I had a portfolio of photographs, and

offers to model quickly came, to walk the runways to techno pop music and strobe lights, or go on shoots with filtered lights, oiled skin, and giant wind machines. But the lure of the money called to me like a siren's song. Top models made $200 an hour, $1,500 a day. A smile, some buffed biceps, and a ripped abdomen were worth in one day almost as much as four years in Tufts had been to the NCAA when it was giving out its post-graduate scholarship awards. If I modeled, and modeled well, I could pay for law school and put some aside. The National Guard was already willing to allow me a short reprieve. The exposure from the *Cosmopolitan* pick had turned out to be helpful to the Guard as well. All I had to do was keep up with my monthly training obligations. I worked out a deal with BC Law School: I took classes at Cardozo Law School in Greenwich Village and was granted a temporary leave of absence. I thought I had everything figured out. Even Ruth, who was still patiently waiting.

One of the requirements of being the *Cosmo* guy was to declare that I was single and to deny having any kind of girlfriend. It was part of the unofficial agreement. Every day, in every interview, I had to deny Ruth. "Yeah, I'm single. Not seeing anyone serious." But of course Ruth was there, and it was an unbelievably hurtful thing. Every time I said it, I knew that I hurt her, even though she almost never said anything. Still, I felt awful. It was a constant cycle of pretending and prevaricating. She couldn't be there with me when I went out. I had to do everything alone. We barely made it through the summer and into the fall.

Ruth was the one who lent me her bicycle so that I could get around Manhattan and over to my law classes. I had started out in cabs, but they cost a fortune, and I hated the subway, hated walking down the trash-strewn stairs to the station, brushing past homeless men and women, their faces blackened from the sun and the wind, shuffling on swollen, gargantuan feet, wearing layers of coats even in the summertime, their things in ripped paper bags or in rusty shopping carts. Even more, I hated the smell as I waited for the train, the fetid heat of the platform, the stench of old urine, and the thunderous noise of the cars sailing past. I had to stay aboveground. Ruth suggested a bicycle, and that was how I moved through the city, weaving around cars, crossing through the park, finding the routes nearer to the two rivers for a bit of a fresh breeze, to combat the lingering smell of the black plastic bags of trash that piled up, massed on the sidewalk. The bike stayed with me, but Ruth and I drifted apart. I was being sucked into a glittering world that wanted only good-looking, unattached men. I didn't know how to navigate that landscape and keep Ruth, or if it would have been possible to do both. Without *Cosmo*, we might have built a future together. Instead, she went overseas to travel with a friend, and things ended because it was easier and also in part because I let them.

It was a mistake that I had to make, to ensure that I never made it again.

I needed the bike, even when the weather got cold, because while the goal was money, at the start I wasn't making anything. I did a bunch of high-end couture runway shows, for which I was paid in clothes, and most of it was completely over-the-top. In one, I walked up and down in a bathing suit wearing a giant mask over my face, while a flock of topless women paraded around me. I had a fabulous wardrobe, but barely $100 to my name. I became practiced in the art of free happy hours, of knowing the places I could duck into starting at about 5 p.m., where if I ordered a Coke, I had free rein at the steaming bins on the buffet table—mini quiches, pigs in blankets, spinach dip, whatever slightly stale, mass-produced food could be set out with little Sterno candles blazing underneath to keep it lukewarm. In that way, many nights, I ate dinner for $1.99.

I was living in a shared rental on Sullivan Street, where I dined on store-brand macaroni and cheese and spaghetti from cans. One morning, I woke up covered in cockroaches making the trek across the blanket and down my legs and up my arms. I might have given it all up, but I got an offer to become the face and also, inevitably, the backside of Jordache jeans.

Chapter Twelve

JUMPING OUT OF PLANES

Modeling in the early 1980s was all about the "go-see." Go-sees were interviews where hopeful candidates crowded into waiting rooms, trying to look completely nonchalant, while clutching their glossy-photo portfolios and waiting to be called. Inside, casting directors looked through the images and said things like "Take off your shirt" or "Put this on." It was the heyday of *Interview* and *GQ* magazines, where men strutted around with their jackets slung over their shoulders and the powerhouse agencies like Wilhelmina, Elite, and Ford ran the business, deciding who was the next big thing and the next new face. I started with the Sue Charney agency, and within a few months, I had signed with Wilhelmina. The modeling business back then was a funnel system; all the young, good-looking hopefuls were dropped into the same vast container and only a few were pushed through the narrow stem to come out on the other side. I usually ended up at go-sees with the same group of men. Most of us were regular guys, usually athletes, in shape, with decent looks. Sometimes the wheel rotated in our favor; sometimes it didn't.

Right around the start of 1983, I went on a go-see for Jordache jeans. Jordache had vaulted to superstardom in the designer jeans market in 1979 by producing a commercial with an apparently topless woman galloping on a horse amid breaking waves. It was too hot for network television, but in the era of Brooke Shields in *Pretty Baby*, that only made it more attractive to potential customers. By the early 1980s, the three names stitched across most fashionable rear ends were Calvin Klein, Gloria Vanderbilt, and Jordache. Now Jordache was looking for a new face for its ads, on TV and in print. The contract was for $20,000 and I made the cut. Within weeks, I had a billboard looming in Times Square, in full view of the discount Broadway tickets booth, the glinting neon lights, and the parting sea of taxicabs. And I had filmed a commercial. I was rich, or as rich as I had ever been. I left the cockroaches of lower Manhattan and Greenwich Village behind and started living in a hotel suite at Fifty-eighth and Park with a bunch of actors and models. Each day, unseen maids came in and tidied up our beds and picked up our towels off the floor.

I took acting lessons—because all models really wanted to get out of modeling and into Broadway or film roles—went to my law classes, and ran around so that I could be seen at the same clubs as Christie Brinkley and Linda Evangelista. The 1990 ad campaign for Canon cameras, "Image is everything," was about eight years too late for the New York modeling scene. When I was sent on a go-see for a golf company, I B.S.-ed my way into convincing them that I could actually play golf and spent a weekend in Bermuda, wearing golf clothes and shanking the ball out of the sand traps.

One of my New York friends had thrown a birthday party for me at the Underground, another dance club, where he hired strippers dressed as cowgirls, which the other guests enjoyed far more than me. When my high school buddies visited, we were waved through the rope lines into whatever nightspot we wanted. But they also shook their heads and rolled their eyes at the life I had joined. They openly mocked the pretense of it all, where inside the clubs, neon strobe lights bathed the smoky rooms in a strange, pulsing light and everyone had to shout just to be heard. While the crowd around us ordered imports, like Heineken or Corona, or shots of tequila where each drinker claimed to have gotten the worm, my buddies stuck to bottles of Bud and said, "This is not real life, Scott."

I knew that too. For me, modeling was a business, a way to get some financial security and to hedge my bets. I had already done the series of calculations in my head—if the modeling didn't work out, I had law school; if law school didn't work out, I had the Army. I had two or three backup plans going at once, so I would never be dependent on any one thing.

And I had already discovered the seamy side, like the squeegee guys who lurked in the shadows after midnight, waiting to jump out and spray the windshields of the cars that cruised Manhattan whenever they were idling at a red light. Right after I arrived in New York, I had joined the Screen Actors Guild and the American Federation of Television and Radio Artists (AFTRA) so I could get appearance fees when I went on TV as the *Cosmo* guy. Everybody seemed to want a piece of me, and I in turn trusted almost no one. I had months where I was rich and months where I had nothing except the wardrobe that the fashion houses gave to me in trade. I took jobs in other places, including a couple of shows in Philadelphia. After one, I met a local television personality named Bonnie. She invited me to the symphony. I said sure and after a few more visits I was splitting my time between Philly and New York, with trips each month up to Boston to train with the Guard.

It was a treadmill of always looking good, always going to the right places, and going on the go-sees. I was becoming hard and cautious. At the start of the summer, I got a call from my professor of military science. I had been given a training deferment the previous year—the Army had even used me to model for one of its recruitment ads. But now, I had to go to advanced training in North Carolina or be bounced from the program. I also got word that I had to return to Boston College Law School, drop out, or transfer to make space for someone else. I believed wholeheartedly in the maxim "If you start something, you finish it." You never ignore your obligations. I asked for a couple of days to think about it.

The next night, I had an invite to a party for Christie Brinkley at Studio 54. I walked in and I was once again the *Cosmo* guy, the guy on the posters and the beefcake wall calendar, the guy whose face zipped across the screen for Jordache denim. The last thing I remember is talking to a New York socialite who was a fixture at the club, with her permanent tan and iridescent eye shadow, oversize earrings, and expensive, low-cut dress. She had gotten me a beer. I had only three beers all evening. The next morning, I woke up in a bedroom and didn't know where I was. Slowly, it dawned on me that I was in the socialite's apartment. There were pictures of presidents on the walls and her assistant was flitting around the rooms. But I had no memory of how I got there, no memory of anything. The whole night was erased, as if someone had slipped something into my drink, which is probably what happened. I dressed and left, walking down Madison Avenue, staring back at my reflection in the window glass, saying, "Who the hell am I?" I looked at myself and did not like the person I was becoming.

That morning, I called my military science professor and asked him if the opportunity to go to advanced camp was still open. He said yes. I went back to Boston, I completed all of my tests, and two weeks later, I was standing on the parade grounds of Fort Bragg, being yelled at by drill sergeants. It was the best decision I ever made.

At the start, my reputation at Fort Bragg preceded me. I was the *Cosmo* guy, which is what my fellow cadets and the instructors called me, when they weren't calling me "pretty boy" and "sissy boy" and "twinkle toes" or whatever else came to mind. But I was determined. I would win the races, do the best in the physical training test, and come out on top in land navigation, or "land nav." And there was a kind of grudging, mutual respect that we developed in those first few days and weeks. No matter how much they ribbed me, we were all here doing the same stuff, and I wasn't trying to coast on any reputation. No other *Cosmo* guy had signed up for advanced U.S. military training. At Fort Bragg. In Fayetteville, North Carolina. In July.

Nothing drove home where we were and what we were doing like land nav, which was also the most nerve-racking of all our assignments. I could always do the athletic stuff unless I got hurt. The classroom requirements were never a problem. But land nav was pass/fail or, more appropriately, do-or-die. Fail land nav and you're done, out of the program. For each navigation exercise, every cadet would be given a map, coordinates, and a compass, and told to go find this spot. And it was timed.

The start was the most critical thing, orienting the compass and the map to make sure that I was going in the right direction. The courses were miles long and I would literally be running to make sure that I reached the end in time. Head southeast rather than southwest, and I would destroy my chances from step one. And in land nav, I was alone. There was no one running the obstacle course beside me, no next group of people doing sit-ups or push-ups, no one scribbling with a pencil at the desk to my left or my right. I was dripping sweat in the bushes by myself.

I came out of one cluster of bushes covered in ticks, so I had to stop for a few minutes to pull them off before they burrowed under my clothes and into my skin. Every brown, hard-shelled tick added to the stress because another second was being lost to the clock. The outdoor courses were covered in ticks and chiggers, just waiting to overrun my bare neck or hands or arms; no-see-ums, with their quick bites that left red welts; and long, slithery snakes. The weather was hot and humid; the air was so saturated with water that it felt almost liquid. Even the rain didn't help. Afterward, it was only wetter, hotter, and stickier.

I remember one land nav round right at the end of advanced training. We had done PT and our firing exercises. I set off and I couldn't find the first coordinate, so I ran back to the starting point, reoriented myself, and set off again. I ran to the first spot, then the second, then the third. I finally found my groove, spot and go, spot and go. I knew that one running step equaled about three walking steps. Usually I'd try to pace myself, walking and running, but this time, because I had started late, I kept running, weaving into and out of trees, crunching over the twigs and dry sticks scattered on the ground. And I finished before anyone else. I was on the bus with the lieutenant being checked for ticks when the rest of the cadets came in. The ticks were so numerous that everyone had to take off his or her uniform. The ones who were covered from head to toe had to also drop their drawers, bend over, and be checked. It built a strange camaraderie too, each man or woman depending on his or her buddies or officers so as not to get scavenged by tiny bloodsuckers.

Right before advanced camp ended, I earned one of the coveted airborne training slots and was invited to stay on for a chance to earn my wings. The airborne training was being held not at Fort Benning, the traditional training site for airborne school, but at Fort Bragg, and the usual three weeks of training were condensed into two. Everything was accelerated. I was eager to do it. Back in Philly, Bonnie thought that I had lost my mind.

The thing that gets you the first few times in the plane is not the pitch and roll when the pilots fly "map of the earth"; it is the smell of the engine fumes, the burning oil and gas that enter the cabin through the open door in one onward rush. The smell is not just a smell; it is a collection of thick exhaust particles that coat the inside of your nose and the back of your mouth. So there will be a bunch of guys, lined up in rows along the inside of the plane, in full gear with their helmets, their clips, and their chutes, bouncing up and down, smelling this noxious smell, in the full-on summer heat, ears ringing from the noise. And at some point, one of them will throw up. The trainers prepared all of us for this. We were all handed paper airsick bags. Once they were used, they were to be deposited inside our shirts. This meant only one thing: they would explode upon impact.

I was determined that this would not happen to me. I was at the back of the line, one of the last jumpers. The first guy who puked was up toward the front, closer to the door, nearer to the wings and the whir of the engine. But the puking moved toward the back of the line like a reflex. One guy vomited, then another, then another after that. The smell was the trigger. I was five minutes from my jump when I puked into the paper bag. And then I did exactly as I had been told. I put it in my shirt, next to my chest. And when I landed, upon the force of impact, the bag burst, covering me in vomit.

We did day jumps and night jumps. We learned how to hit the ground. The best way to do it is to land on your feet with your knees bent and then roll to your side. When you do it right, it feels almost perfect, not much different from getting out of bed. The worst jumps are when you land with your feet first, then your head, and then your butt. Anyone with that landing is basically smacking his or her head against the ground upon impact. Even with a helmet, the force of the hit leaves you dazed. And the first thing you are supposed to do when you hit the ground is unhook your parachute. But if it's windy and you're dazed, you can't unhook the chute. The wind will lift up the silks and basically reinflate the parachute, and you'll be dragged until something forces it all to come to a stop. Until that moment, a stunned jumper is trapped by the massive, billowing ball of material. I had a feet-head-butt landing once, and I learned my lesson; I hoped never to have it again.

Of all our maneuvers, the night jumps were by far the scariest. When you jump out of a plane at night, you don't know when you are going to hit. You are falling through the black sky with very little idea of how far you are from the ground. The only way to gauge your distance is to yell, and then the echo from the yell bounces off the earth. The faster the echo bounces back, the closer you are to the ground. Then you start looking for the tree line, the darker shadows against the very dark sky. When you see trees ahead, you know you are going to hit. It's one, two, three, here it comes, here it comes, boom: smack against the ground or snagged on a pointy rush of branches, if you are particularly unlucky that night.

When my group of cadets passed Airborne School, we earned our blood wings—our senior officers and trainers took our pins and jammed them straight through our shirts, into our flesh.

I kept in constant contact with Bonnie during my training, and she even came down to visit me one weekend. I met her in my uniform; she showed up looking like a fashion model; and we drove around trying to get a place to stay for the weekend. Because we were an unmarried couple, it took us about eleven tries before we found a place in North Carolina that would rent to us. It was a shock to me, being from Boston. There, hotels would take your money without a second glance. But not here. The futile search for a place was a bit like a metaphor for our relationship. That was over too. She had wanted a model, not a guy in the military who was going back to law school. Our lives were moving in two different directions. I left Fort Bragg and headed back to Boston to finish law school. One local newspaper actually ran a piece reporting that the *Cosmo* guy had returned. I kept up my classes and also continued to model on the side. I had enough money now for a down payment on a condo and a car. I kept doing my National Guard training and finished ROTC to become a commissioned officer. I still had all my options to keep open.

A few months after the *Cosmo* spread appeared, I did an interview with Marion Christy for the *Boston Globe*. The *Cosmo* experience had left me emboldened, and I was honest with her, more honest than I probably had ever been. During the interview, I talked about my father's absence from my life, and how he wasn't around while I was growing up. She put all of it in her piece.

At the same time, I had also been trying to reconnect with my siblings, to see them and get to know them without some of the baggage and resentment that had polluted our growing up. One Saturday when I was in Boston, I decided to head up to Peabody to watch my half brother, Bruce, play baseball. My dad had also showed up to watch. At this point, it was rare for us to speak with any kind of routine. He had split from this third wife, and was now getting ready to marry for a fourth time, with Peggy, the ex-wife of one of his former business partners who had passed away.

That day, he came moping up to me and said, "I read what you said in the newspaper." Most of his friends had read it too. It was all true, but it stung him. He started to talk about the games he had come to and how it was really my mother who had kept him away from me at home. I didn't want to hear it again. I practically had to sue him to get the equivalent of petty cash to feed myself when I was playing varsity basketball at one of the top academic schools in New England. I didn't want to go back over the past. I told him point blank, "I'm sick and tired of the bullshit. Either you're in my life or you're not. Let me know what you want to do." We stood there, looking at each other, and he said, "I'd like to try to get back into your life."

It took years, but we both made an effort to reconcile. We started with simply trying to call each other more. Slowly, we started to get together, to meet for dinner. He quickly learned not to make me feel guilty about my mom, and when she complained to me about my father, I told her that I didn't want to listen. Every few months, we made another tentative step. Eventually, my dad invited me out to Newburyport, where he would take me bar crawling up the stretch of road toward the New Hampshire shoreline. We'd wind our way from one windowless dive to the next, where we drank beer and listened to out-of-date music in smoky, wood-paneled rooms, at slightly sticky counters worn smooth from years of rubbing down, and inside each place, he proudly pointed me out to his friends. We would talk more and try to find some common ground.

As the first twenty-five years of my life drew to a close, I was finally getting to know my old man. And I really believe that it never would have happened without *Cosmopolitan* magazine. If he hadn't read that later article in the *Globe*, hadn't seen those words laid out on the page, we might be exactly where we had always been: quasi-strangers who happened to share a last name.

I made one mistake in law school, a smart and very pretty girl, whom I was infatuated with but whom none of my close friends liked. We got engaged anyway, until I found her coming home from a weekend with a guy who was her old boss and happened to have been her boyfriend as well. After that discovery and with law school ending, I took off to Jamaica to lie on a beach and lick my wounds. There, I met a beautiful French-Canadian named José from Montreal. She spoke limited English, and about the only French I could decipher was what had come from my increasingly long-ago Latin, but after the drama with my ex-fiancée, the language barrier was perfect. After our time in Jamaica, we stayed in touch. I headed up to Montreal to see her; she came to Boston to see me.

In the early spring of 1985, I was off to Fort Benning, Georgia, for more weeks of infantry training. At Fort Benning we were up every day before 4 a.m. By 4, we would be on the course. Running, jumping, sprinting, push-ups, sit-ups—we were deep into all of them by 4:30 a.m. The sun was just coming up overhead as we finished. Then we had marksman training, classroom time, athletic competitions, and maneuvers in the field. One of our biggest field exercises was for eight days, with basically no sleep, maybe just a couple of hours a night. It's a test, not simply a field test, but a test of who's going to fail and who's going to succeed, who's going to crumble under the pressure or rise to the occasion.

The first day went great, and so did the second and the third. Each day, I waited to see which night in the field would be mine to lead. I was assigned to lead the last night, the night before the troops were set to go in, the night when all the men hit the wall. It's cold at night, even in Georgia in the spring. In the woods, the temperature falls quickly. I was handed a scenario, with a location of the enemy. We had to move on the enemy and move in quietly by 6 a.m., or I would fail. We had night-vision goggles strapped to our heads to see in the darkness. I got everyone moving at dusk, but once there was no longer any trace of sun in the sky, everyone in the unit just kind of died. Men sat down on logs, leaned against trees, or lay down on the ground and just curled up. They might as well have gotten a book or a clicker and pretended the nearest tree trunk was the back of a couch. They wouldn't move. The clock was ticking, and I had to go around smacking people, yelling at them to get up. We ultimately made it through, but only by the skin of our teeth. It was the low point of my entire military summer training period.

But I stuck it out. I knew some of it was payback. I was a Boston College Law School grad about to take the bar, I was a model—you could still pick up a magazine and see my face or catch me in a commercial—and here I was, in the infantry, in the woods. I had driven down in a nice car, a gold Dodge Daytona, and I had a Canadian girlfriend, so when other guys were going on leave to cruise the bars in Columbus, Georgia, I had the cash to fly up to Montreal for the weekend. I was a little older than most of the other guys too, and yet I was still winning a lot of the athletic competitions. They should have busted my chops. If the stories had been reversed, I would have been sorely tempted to do the same.

We were all officers in the infantry course, and nearly all of the men here were commissioned through ROTC. Some were being trained to go on active duty, to lead their soldiers into battle in some other part of the world. Others would return home to serve in the National Guard or Reserves. On many of our exercises, I learned a tremendous amount about tactics, logistics, battle skills, leadership, POW awareness, and even first aid. We spent hours and hours in the field, honing our skills to become infantry officers. When I finished my training, I was assigned to a headquarters company in Boston. I held the rank of second lieutenant and was now the number two officer, behind the company commander, Captain Valente. We had a great group. Whenever the Regular Army officers came to evaluate us on our field training, we always did well. Too well. We were one of the higher-ranked companies, and Captain Valente knew that I was a kind of hard-charging guy, and he could delegate a lot of the field training setup and details to me. It was not in his interest to see me promoted out of his company into something else.

My way out came when the Guard downsized. I was in the Yankee Division, about ten thousand to twelve thousand people, but in the downsizing, we were cut from a division to a brigade. The Guard did away with the headquarters company and the infantry, so I had to change branches. I became a quartermaster officer. I went to quartermaster advanced school one summer and came back as a company commander, personally responsible for millions of dollars of equipment and for all our training. But by then, I was married with two kids. It was becoming less enticing to spend weekends in the woods. I looked into military intelligence, and then realized that the smartest thing I could do would be to start doing in the military what I did in civilian life, work as a lawyer. So I applied to become a judge advocate general (JAG) legal officer. I was accepted and did my training down at the University of Virginia in Charlottesville, surrounded by a treasure trove of history. I had been promoted to a captain in the Guard and served primarily as a defense lawyer, which meant that when people in the military did stupid stuff, I was the guy they called.

My first-ever rotation as a JAG lawyer was in Fort Drum, in the northwestern corner of upstate New York. Not long after I had arrived, I got a call at 3 a.m. and heard, “Hi, sir, we’ve got a couple of guys who jumped off into the water and were trying to swim to Canada.” I was sound asleep at the Fort Drum Inn. I said, “Yeah, right,” and then the phone rang again. The guys were tankers, meaning they drove around in tanks. Tanks are hot; the men in a tank sweat, and when they get out, they're often dehydrated—a condition that makes their alcohol tolerance go down. When these two guys got off maneuvers, they headed up to Alexandria Bay, on the St. Lawrence River, and started drinking beer at a place with a power hour and all the beer and hard liquor you can drink for about $10 a head. They went inside, drank, then went out back to pee on the lawn and saw a couple of three-story paddle boat pleasure cruisers. In their unit, one soldier's nickname was "Pirate"; the other was called something like "Long John Silver," and he said, "Hey, Pirate, let's go steal the boat."

They untied the moorings, got into the cabin, and tried to switch the boat on. But the engine wouldn't turn over. They were a kill switch away from getting power. But without the moorings, the boat had drifted away from the dock. So now they had an unmoored boat that they couldn't steer and it was moving down the river. They banged into other docks and hit boats, until they finally ran aground. By that point, they were sober enough to realize that they were in big trouble. They saw the cops coming and they jumped overboard with the idea that they would swim across to Canada. But they were still too drunk to do that. The cops picked them up, booked them, and took them to the county jail in Watertown, New York; and because they were Army guys, they called the base, which then called me. CNN cameras were already on the scene and all the local stations had turned up to cover the aftermath; an event like this was big news.

I was working on the case with Major Karen McNutt, a great lawyer and a mentor of mine, and soon afterward, I headed over to Watertown to consult with the two soldiers in the jail. They were brought into a holding room wearing orange jumpsuits. And they were laughing. "Oh, how ya doing, sir?" they said when they saw me, completely cracking up. And I said, "I don't know what the hell you guys are laughing about. This is my first case as a JAG." That was technically true. Of course, I had been a practicing attorney for about ten years, and this was an outside case, not a military case, so I knew what I was doing. But that got their attention. All the color drained from their faces and suddenly they were very focused. Both guys were Desert Storm veterans, one was highly decorated, and in the end, for their military punishment, they had their ranks reduced and were heavily penalized.

The civilian case against them had a happier ending. When I went over to the Watertown courthouse one evening to meet with the judge, I saw a guy going in wearing a T-shirt from NAPA Auto Parts. He looked like a nice guy and I said hi. Ten minutes later, I'm meeting with the judge, and it was the NAPA Auto Parts guy. He may have even been a judge part-time. We began talking, and he asked me where I was headed next, and I told him I was going to Newark, New York (a small town in the western part of the state), to visit my uncle Tom and my aunt Linda, my dad's sister. I had seen her every once in a while since I was about ten. The judge asked me for Tom's last name, and I said, "McHugh." Tom McHugh was a detective and about seven feet tall, and the judge knew him. We talked some more, and he dismissed the case against the two soldiers. They were only required to pay a fine. "They sound like young guys who made a really stupid mistake," he told me. But, he added, "Get them out of here and don't have them come back."

I had many more military cases after that, including hundreds of separation boards for drug use, where the first thing I tried to ascertain was whether the accused had ever been high and around or, God forbid, working with a loaded weapon. But I also remembered Judge Zoll, and I investigated whether the guy or woman sitting in front of me deserved a second chance. I got the reputation as the lawyer to seek out if there was a problem, because I always tried to go the extra mile for my clients, the soldiers. So I was very much surprised when I had one soldier come up to me, a sergeant first class from Massachusetts who had been in the service for seventeen years. And he began by saying, "I did it. Kick me out. I don't care." I responded, "You've been in seventeen years. I'm not just going to kick you out. Instead, why don't you tell me your story?" So he began.

He'd been married for most of his time in the Guard, but his wife had left him to join a cult, had shaved her head, the whole thing. Then his house burned down, and he lost everything. Then his daughter was raped. All within a span of two months. Then he was diagnosed with prostate cancer. He was beyond distraught and beyond caring. He went out drinking with his friends. Someone passed around a joint, and he thought, "Why not, who cares?" I listened to him and to his despair, and I worried that if the Guard—the last thing that he had to hold on to—kicked him out, the guy might actually kill himself. He ended his story by saying, "Do whatever you want. I don't care."

I told him that the only thing I wanted him to do was to tell his story before the drug board—five people, two officers and three enlisted men or women. By the time the sergeant had finished speaking at his hearing, all the board members were either crying or trying to hold back their tears. Then I made my argument. I said the National Guard talks about how we are a family, how we watch over our men and women when they go away. Well, here's a guy who has been through some rough things, any one of which no one should have to go through in a lifetime. He's had all of them happen to him in a matter of months. He obviously needs our help, and we're going to kick him out? Are you kidding me? The board voted to maintain him, he got help, and he finished his career in the Guard and retired.

Some of the most difficult days I had were the mobilization exercises: I had to prepare men and women, some young and some no longer young, to go overseas, and after 9/11, to go overseas to war. I've been to Paraguay and Kazakhstan for important duty, but I've never been mobilized myself for extended deployment, at home or overseas. For myself, and for other of my friends in the Guard, there's a feeling of somehow not doing our part because we have not been called to extended active duty. For years, I've wished that I too could go over and serve, but, like all soldiers, I go where I am ordered. I do know that the classes I've taught on how to legally defend our U.S. soldiers or on how to treat POWs matter to our men and women in uniform who are going into the field.

It's ironic, but today when we fight as a nation, military lawyers are some of the first ones to go into a war zone and the last ones out. When we go into countries to stage our troops, we need lawyers to establish contracts for buying food and buying gas, to understand and follow the host country's rules and regulations, its customs and courtesies—even something as basic as what type of photos married couples can send each other. In the United States, a wife can send a husband a naked picture of herself, but in some host countries, that would be illegal. Lawyers have to familiarize soldiers with the rules, the customs, and the courtesies. At the end of a conflict, it's lawyers who finalize any reparations, conclude any contracts, do all the things that come with being a twenty-first-century army. Military lawyers are also responsible for making sure that if there are any problems or infractions, our soldiers are treated under our laws, not under another country's laws.

I was one of the soldier-lawyers who stayed home to help our men and women prepare for war. I was, for all intents and purposes, one cog in a many-geared machine. Before Guardsmen and Guardswomen are sent into today's combat zones, lawyers like me check their wills, powers of attorney, and other legal matters. It's the lawyer's job to review all their documentation in case they are killed or wounded. I was working on that type of legal intake one afternoon when a big, muscular guy, maybe six foot five or six, came in, the kind of guy who needs a specially made uniform because his arms are so massive. He had muscles on his shoulders, his neck, even his elbows. He looked like a trainer or a bodyguard, in great shape, with a flattop. And he looked like the type of guy who, if he needed to, could rip your head off.

At first, as I was starting on the list of questions, he seemed kind of cocky. But there was a bit of an emotional air about him, just under the surface. To my mind, something was clearly wrong. I asked him if he was married, if he had any kids. And he started to tear up, small tears running down his cheeks. He told me that he had just gotten married, just had a kid. He said, "I'll go through that wall. I'll kill anybody. You tell me to jump, I'll do whatever you want. But I'm afraid if I die, my kid will never know me, and if I go away, my wife is going to leave me and I will never see my kid." And he kept right on crying. There are sections in manuals that instruct us on what to do in these types of situations, but I simply grabbed his hand, looked him right in the eye, and said, "What would you like me to do?" He said, "I don't know." I went straight to the unit chaplain and to the appropriate support groups. I told them what was going on, and we made sure that the soldier remained in constant contact with his wife and that the family services groups were involved with his wife and baby during his full twelve-month deployment, that they would help her with whatever she needed, even if it was a babysitter.

Twelve months later, when he came back, he specifically sought me out to tell me that he and his wife were "solid," and their baby was solid. And he wanted to say thanks. I remember him picking me up off the floor and the resounding crack that my back made as he hugged me and said, "Thank you."

Chapter Thirteen

GAIL AND MR. MOM

I met my wife over a napkin. I was home from Fort Benning after finishing my Infantry Officer Basic course, prepping for the bar exam, doing some modeling and a little house painting for extra cash, when my buddy Seth Greenberg called. I had been introduced to Seth when he sponsored a Richard Gere look-alike contest at a local Boston club while I was at Tufts. Seth was very handsome, with dark good looks; women flocked to him; men respected him; and he was making a small fortune in the Boston-area party business. He began in college by throwing parties in local venues and asking for a piece of the door—the entry fee—and the bar. He got people to turn out and the more guests who came, the more he got paid. The Richard Gere look-alike contest was another one of his ideas, except he had planned on entering it himself and had also planned on winning. Unfortunately for Seth, I saw it advertised and entered it too. And I won. But from that night, Seth and I became fast friends. We went out together constantly, and I helped him throw some of his parties, for my own cut of the door and the bar. By the spring of 1985, he was a part owner of a club called the Paradise, which was being used one evening to film a Miller Beer commercial. "There are going to be some hot-looking women here," he told me. "Come on down." So I did.

I walked in with my field-maneuvers tan and the remains of my army buzz cut and started chatting with some of the people hanging around. And then I saw her. Gail had long brown hair, pouty lips, and a face that was not just beautiful—it was mesmerizing. In a room of gorgeous people with perfect faces, teeth, and hair, for me, she was the most perfect of them all. I watched her all throughout the shoot, even when she took a break and went over to sit down and eat. She looked flawless there too, until the moment when she spilled something on herself. She didn't notice, but I did, and I saw my chance. I rushed over with a napkin and handed it to her.

That was all the opening I needed. We made some small talk and I tried to get her phone number, but she wouldn't give it to me. Instead, I persuaded her to take mine.

In Boston, Gail was a pretty well-known model, who was working to put herself through the broadcast journalism program at Emerson College. She also had a boyfriend who was in the news business. People told me they were somewhat serious. We were thrown together a few times in the weeks after that, professionally, working at a couple of modeling jobs. In our first shoot, we had to walk together down the beach, arm in arm, looking out at the sand from opposite directions, our eyes never meeting. But in the photo, you can see and feel the chemistry.

Our paths crossed briefly again when we met outside the Massachusetts labor board, trying to collect on the bond of our modeling agents, both of whom had stiffed us. Mine took a whopping $50,000 of my paychecks and used the cash for plastic surgery and to support his drug habit. When I discovered how much he had stolen from me, I confronted him, grabbed my portfolio and composite cards, and walked across the street into the offices of Maggie Trichon, one of New England's premier agents. I told her, "My last agent screwed me, I'm with you now." And I was, whether she wanted me or not. After that, Maggie had me working almost every day, even as a hand model—I was hockey great Bobby Orr's hands when he filmed a BayBank ATM commercial. I stayed crouched down behind him and then raised my hands when it was time for Bobby to touch the electronic keypad. Basketball players' hands take far less of a beating than hockey players'. As time went on, Maggie trusted me enough to ask me to act as her lawyer as well.

Weeks later, on a warm Friday night, I was home in my mom's last house in Wakefield, sitting in my tiny room under the eaves, my bed a mattress on the floor. I was reading. I was renting out my one-bedroom condo to make some extra cash, and I had moved back to my mom's home with a closet-size bedroom while I worked to get the next phase of my life started. Suddenly the phone rang. It was Gail, telling me, "My girlfriend stood me up. You want to go out?" She was living in Watertown, twenty minutes away. I said, "Sure." It was nearing summer, and I raced up and changed into what I thought was my coolest outfit, a pair of long salmon-colored shorts in butter-soft leather that I had gotten as payment for walking the runway in New York. They were tagged as $1,500 shorts. I paired them with white sandals and a light, pastel striped shirt. I thought I looked great, something like *Miami Vice* meets Ralph Lauren, perfect for an evening in the middle of the 1980s. When I showed up behind the wheel of my gold Dodge Daytona wearing salmon walking shorts, Gail, knowing that I was a model, thought it was likely to be an early evening for her. She was not exactly blown away. But I was.

She came to the door in a thigh-high beige dress with a cutaway back. There was only one place to take a woman like that, the Top of the Hub, Boston's equivalent to the now vanished Windows on the World atop the World Trade Center in New York City. It is a wraparound restaurant perched on Boston's towering glass Prudential building, with views over the entire city and the Charles River, and it was the fanciest place I knew. We ordered drinks and appetizers and sat and talked. Like my parents, Gail's parents were divorced. She had four sisters, the oldest of whom had suffered a devastating stroke brought on in part by birth-control pills when she was just eighteen. The stroke had left her totally paralyzed on her right side. As the third-youngest girl, Gail had been on her own since she was seventeen; after the separation and divorce, both of her parents had moved out of their house, leaving their children behind to basically finish raising themselves. Gail had to complete high school, put herself through college, and make her way almost entirely on her own. I understood her life, not just superficially, but as closely as if it had been mine. And from the start, I admired her for it.

I also learned that it wasn't her girlfriend who had stood her up, but her boyfriend. And I've never been more grateful in my life. By the end of the evening, I was perilously close to love.

We dated for a few months, even though we both knew that we had something very special from the beginning. After Gail, I rarely had time to go out with Seth in the evenings. She was the only person with whom I wanted to spend my free time. Not only was she gorgeous—she still is—she was and is funny and smart. She has a husky, deep laugh that is to me one of the most joyful sounds in the world. Her mind is constantly working in different directions. She is fascinated by the people she meets, and she is uniquely giving and warm. As a twenty-five-year-old guy, I was captivated by her face and her body, but a quarter of a century later, what I find most beautiful is her heart. She is the person I look for first whenever I enter a room. I did it then, and I do it now. Gail is the half that makes me whole.

On a warm summer afternoon, we were in Providence, Rhode Island, where I had a modeling job. We were crossing a busy main street right in front of a new building that was going up, and Gail got down on her knee in the middle of the pavement and asked me to marry her. I started to laugh, not quite sure if she was serious. Then I looked in her eyes and realized that she was. I kept telling her to get up and get out of the street, but she said, "No, I'm not getting up until you say that you'll marry me." All around us, cars were honking, peoplc were shouting, but Gail would not budge. The construction workers on the scaffolding had paused to watch, light glinting off their hard hats, and they were whistling and hollering, "Do it. Do it. Do it." "Say yes. Say yes!" And I did. I said, "Yes. Yes. Yes. Now get up!" We stood in the middle of the street, arms wrapped around each other, kissing, as cars whizzed by, until one of the construction guys yelled, "Get a room!" Then we walked off to the modeling job.

I designed Gail's engagement ring myself; one of the stones in it was a gift from my grandmother and my mom. And when I put it on, I asked her to marry me as well.

The night that I gave her the ring, Gail was preparing to leave for her first full-time journalism job, a dream job for anyone starting out in television reporting: to be the lead anchor at WNCT-TV. In Greenville, North Carolina. I was just trying to get a law practice started in Boston, taking on secondary cases that came into a Boston law firm. I was working with an attorney named John Brazilian, who shared office space with the legendary F. Lee Bailey and other attorneys. I couldn't move to North Carolina, but I wouldn't ask her to give up her career either. If those dark apartments on Albion Street, Broadway, and Salem, or those years of rolling my things up in a blanket and moving from house to house, had taught me anything, it was that all women, if they want, should have the security and identity of their own job and their own career. I would never ask my wife to give up her dreams for me. And two of the things I loved about Gail were her drive and her determination. North Carolina it was.

I became an expert on the flight schedules and the transfers to twin-engine turboprops, skimming through the air on weekends between Boston and Greenville. I would gaze out at the stacked formations of clouds plying their own way up the East Coast and wonder if we would be a commuter marriage and worry about what would give way first.

Because Gail was in North Carolina, I planned our whole wedding in Boston. I found the place, the Tufts chapel, with a party afterward at the university's Inter-Cultural Center. I picked out the caterer and designed the menu. I booked our wedding night at an inn in

Gloucester, Massachusetts, before we flew to our honeymoon getaway in Aruba, where we would stay at the Divi-Divi. All Gail needed to do was to come home, put on her dress, and walk in.

And she almost freaked out. On the morning of our wedding, she was crying and ready to throw up, wondering if she was really able to go through with it. She was a news anchor in the middle of North Carolina who was getting married in Medford, Massachusetts, to a guy who was working as a lawyer in Boston. But she put on her dress. It was July 12, 1986, and we were married inside the Goddard Chapel's one-hundred-year-old stone walls to the steady beat of pouring rain against the stained glass. I remember thinking that it was good luck for a wedding if it rained. Our guests were model friends, photographers, actors, and some of my friends from New York. I also had my military friends, and my high school and college friends. My fifth-grade buddy, Jimmy Healy, was my best man, and I had many other friends in our wedding party. I sent a very special invitation to Brad and Judy Simpson, who came too.

A pastor who was a friend of Gail's married us and we recited our own vows, which we had written. I told her, "Gail, my love, you are my best friend and my love. I pledge to you, in front of God, our friends, and our families: my love and support today, tomorrow, and forever." I envisioned our future life, "growing together, caring for each other, and sharing ourselves, our thoughts and our dreams. To you, I pledge fidelity, sincerity, and honesty." I promised too that I would communicate and "show compassion toward you when compassion is needed." I told Gail, "I love you for your beauty, your intelligence, and just your overall being. I am yours always."

To me, Gail promised, "When you are strong, I will love you; when you are weak, I will love you—forever I will love you despite the hardships we face in this life we've chosen to share together forever. I pledge fidelity, sincerity, and honesty to you, and the openness to be there whenever you need me. I'm yours always." Those words have only come to mean more to me in the intervening years.

Afterward, with Gail glowing in her white dress, and me in my tux, we walked down the campus street under a sea of bobbing umbrellas to our reception. Most people worry about the marriage ceremony, the lifelong commitment, or the vows. For us, the reception was fraught with far more peril.

Divorced parents add a challenge to any adult child's wedding, but we had not one set to contend with, but two, with three remarriages, and then my mom. Our receiving line consisted of Gail, me, her mother and her new husband, my mother, my grandparents, Gail's father and his new wife, and my father and his new wife. The stress level of just shaking hands and accepting congratulations was astronomical. We were both terrified that something would be said and the whole event would dissolve into chaos, a sea of hurt feelings and years of recriminations. We had eleven people in the receiving line, seven of whom we had no way to control. (I didn't worry about my grandparents.)

For the meal, we rejected the idea of a head table, and we let the various family divisions stake out their own encampments, while Gail and I kept circling the room, checking on everyone. It was a series of "Hey, how you doing? Good? OK, see you later." Finally, after a few questions—"Why is your father over there and your mom on the other side over there?"—and some looks and shrugs, we realized that things were getting tense, and we gave up. Gail and I had a glass of champagne and decided to start enjoying ourselves. We left our parents to fend for themselves, and eventually even they ended up having fun. Our guests had a great time, including a few who decided to hang from a second-floor balcony. It was the first wedding reception that the cultural center had ever hosted and I believe it was the last.

We also paid for the entire wedding ourselves, and

we were very grateful for our family and guests who gave us cash as gifts. I think we came away with about $300 after all the bills were paid. That night, driving up to Gloucester, we stopped to buy a pizza because we were starving. In all our circulating around the tables, we had forgotten to eat anything ourselves.

As our wedding day approached, Gail had made the decision to leave her job in North Carolina and take the big gamble of looking for television jobs in the highly competitive markets around Boston. In 1986, for a woman, it was particularly risky to leave your contract over a month early to follow the man you loved. The belief was that commitment to your career was paramount; the heart could and should wait. Gail chose the opposite course, and I loved her all the more for it. We had bought a one-bedroom basement condo in Brighton for about $60,000 with some of my modeling money, and in those first months of our newlywed lives, she grimly worried whether we could afford it. Finally, in late 1986, a job opened up at WLNE in Providence, Rhode Island. We got out a map of Massachusetts and I put my finger in between Providence and Boston, splitting the distance between the two cities. The town I landed on was Wrentham, and that was how we picked where to live. A town of less than ten thousand people, it sat right on the Massachusetts border with Rhode Island and dated back to 1660. It was burned during King Philip's War, the bloody yearlong conflict that raged through the New England colonies, pitting local Indian tribes—led by a Wampanoag Indian named Metacom, who was known as King Philip—against the English colonists and converted Indian Christians. Hundreds of colonists and some three thousand Indians died during the conflict. Wrentham's other major claim to fame was as a town where Helen Keller had lived as an adult. On the north side was Foxboro, Massachusetts, where the New England Patriots football team plays. I sold the basement condo and my other condo, the fixer-upper that I had been renting, and we started looking for a house.

Everything seemed far too expensive for our budget, except for one home, a wood-frame house, in an old Colonial style, sitting close up against the main road, East Street. Its location was a remnant from the days when settlers built their houses right next to whatever carriage or horse path ran nearby, so as not to miss a single passing traveler or to have to go too far to get out when the heavy snows billowed up against their doors, shutting them in. No one ever imagined paved roads and motorcars. The house we found was not a historic house, but it was still close to the main route into downtown. And it was several other things as well. It was on the market as part of a bankruptcy sale, the home of a plumber who had gone bust. The front looked all right, but the backyard was a dumping ground of broken toilets, old tires, rusty pipes, and other assorted plumbing debris. It was so covered with junk that it was hard to see more than a few patches of neglected grass. Amid all the trash, there was a pool, filled with murky, disgusting water and algae growing all along the sides. But it was what we could afford, and Gail and I said, "We'll take it."

I spent months hauling the junk out of the backyard and scraping off the peeling paint to prime it and repaint. I got into the pool with a toothbrush to clean the corroded sides, my legs sopping wet from the brackish water that collected in the concrete after each rain. It took over a year to reclaim the yard and begin to rehab the house, and almost as soon as I had finished, the opportunity to move came again.

Behind us, off East Street, was a beautiful residential neighborhood with two dead-end cul-de-sacs. I used to walk or run or ride my bike through it and think about how I would like to live there. One day, out of the blue, a guy who had a house on a nice block came up to me and said, "I'm getting divorced. Would you like to buy my house?" Gail was working, I was working, we were paying the bills, but we were still stretched to the limit, with no extra money. However, I loved that neighborhood and I said, "Sure." I borrowed and scrimped to put together a down payment on the house. We put our other house on the market, but it was fall, school was starting, the market was slow, and it didn't sell. Now I had two houses and one job, and I was thinking: What the hell am I going to do?

It was 1987, which happened to be the year of the National Football League players strike, and the New England Patriots, who played nearby, were bringing in replacement players for the season. The local paper ran an article about how the replacement players needed housing. I immediately called the stadium, tracked down one of the housing reps, and offered my new house, for three months, until we moved in. The players were paying $200 a night in local hotels. They were happy to have my house. I ended up with five players, a quarterback, and some receivers, all nice guys and very religious. To get the house ready, I raced around borrowing extra beds and furniture from friends. For over a month they paid me rent, and soon after they left, I was able to sell our house, and I even managed to refinance the mortgage on the new house—ultimately paying less for this new house than we had been paying for the old one.

We moved into our new home in the dead of winter, right after a big snowstorm, which fell heavy and icy, with a hard, frozen crust on top. For me, it was a blessing. Since we were literally moving around the corner, I saw no reason to spend the extra money to hire movers, especially since we had no extra money left over to spend. I had to move everything myself, including our washing machine and dryer. I began to push the machines out of our old basement, up the backyard hill, through the fence, and out onto the street. I ran them over the ice-crusted snow like a sled, sliding them a block and a half to our new home. Little did I know that one of our neighbors, Dante Scarnecchia, was watching me. He was, and still is, an assistant coach for the New England Patriots. It was cold—that crisp, clear cold that settles over Massachusetts after a snow—and to keep my body warm, I had put on my heaviest clothes, my army fatigues. Dante stared out his window as some guy in army camouflage pushed laundry equipment up the block, sweat beading on my forehead and breath coming out in giant, cold puffs, moving all of it right into the house next door to his. As he watched me arrive with a fresh load, push it inside, and then set off again, he grew concerned and began to wonder if his new next-door neighbor was some kind of survivalist nut. Because it didn't end with the washer-dryer. After those, I slid the refrigerator and a whole bunch of boxes along the same route. Dante told me that it took him a couple of weeks to get over the sight of my do-it-myself moving operation and to realize that I was just a guy doing it all alone with no help.

Before we moved in, Gail had gotten pregnant. The television station in Providence had no maternity policy. It was trying to force her to take just two weeks of vacation after the birth and then return to work. The station wouldn't even grant her unpaid leave. That did not seem right to us. I ended up representing my wife and getting her eight weeks of vacation to cover her maternity leave once Ayla was born. After that, the station instituted a policy for pregnant employees.

The pregnancy was relatively easy, but the birth was very difficult. Ayla was breech, with the cord wrapped around her neck and in fetal distress; her heart was stopping as the cord pressed down ever tighter on her neck, which had also happened to me when I was born. Gail had to be cut open with an episiotomy and Ayla was pulled out, bruised and cut, with forceps. She came into the world on July 28, 1988, wailing, as if she were saying, "Damn, get me out of here." We chose her name from the book *Clan of the Cave Bear*, by Jean Auel; Ayla was the main character, a woman who was a battler and a fighter, and Ayla has followed her namesake from day one.

The drama didn't end there. Gail's stitches got infected and she developed a blood clot under her arm that landed her back in the hospital for a week and a half and left me home with a newborn. Thankfully, we had a bit of help from Gail's mom and some neighbors, but mostly we were on our own and totally overwhelmed. I remember a blisteringly hot Saturday afternoon in August when Gail had finally come home from the hospital. I was mowing the lawn and she was sitting in the little breezeway at the front of the house with Ayla. And over the noise of the motor, I heard sounds. It sounded like crying and yelling; something seemed to be wrong. I let go of the lawn mower and bolted to the breezeway to find both Ayla and Gail crying and Gail trying to feed her, breast milk spraying everywhere across the six-foot porch, including into Ayla's eyes and face. Gail was sobbing, saying, "I can't do this. I don't know what to do. I'm a terrible mother." It was a sea of tears, milk, sweat, and snot, and I—after all those years in the National Guard and playing basketball—did what any new dad confronted by two hysterical females, one very small and one not as small, would do. I took Ayla's head in the palm of my hand and wedged it right against Gail's breast until she finally latched on and began to nurse. And then I told Gail that she was a great mother and made a really bad joke, as I still like to do. My really bad jokes and some of Gail's have been the balm for many of the trying days in our marriage.

Like many couples with a newborn, we were now trying to juggle not only our lives, but a new baby, a new house, and our jobs. We had to find good babysitters, which was a constant struggle. Some of our first few were terrible, and at least one mistreated Ayla. We didn't know much about finding good care. We worried about Ayla, and we worried about being home, about paying the bills, about how we would do it all and keep it all together. I was twenty-nine years old and I wished that I had someone I could call on for help. Gail and I would constantly say: Where are our parents? Can't our parents come and help for a while? Most of our friends who were new parents had help from their families. We really needed it, and we barely had any. Over time, some of our wonderful neighbors, especially Tom and Sandi Gamelin and Joe and Gerri Pavao, stepped in, but in those first years, we were just muddling through.

Looking back, I'm amazed at our resilience and Gail's composure. Here was a woman who was taking care of a baby with no outside help around the house, and still working at a grueling job in television news. In fact, her work got harder after Ayla's birth. Gail was offered a spot at a bigger station in a bigger media market. But the job was in Hartford, Connecticut. It meant a promotion, but it also meant driving a 218-mile round-trip each day from Wrentham. Gail has never been handed anything; she has worked for every opportunity, every position. She is an incredibly diligent and dedicated journalist. For a year, she drove back and forth to Hartford, filed her stories, and met her deadlines. I tried to pick up the slack at home. I did laundry; I changed diapers; I read and sang to Ayla. I learned to know almost before she did when she was hungry or tired or just needed to be picked up and cuddled.

In those early years, my big outside vice was basketball. Some women's husbands make up excuses to go to bars; I made up excuses to sneak out and play hoop in a couple of local leagues. One of the first things I did at our new house was put up a basketball hoop with a glass backboard in our driveway. For years, I had dreamed of having that pro hoop, and now it was finally mine. I used to wear my basketball uniform under my clothes or stuff it inside my briefcase. I remember one time telling Gail that I had to get to a meeting. She asked me where my wallet was and I said, "I think it's in my briefcase." She popped the bag open and there were my high-top sneakers, shorts, and team uniform shirt. But when we played our games outside in the parks in the summer, she would often come with Ayla, and once Ayla got bigger, I would drag her to my practices myself. I would strap her in her little wheeled bouncy seat or, later, her stroller, and she would scurry back and forth along the gym floor or outdoor court, trying to keep up with us as we ran from one end of the court to the other.

One of our league championship matches was on the actual day of Ayla's christening, and Ayla's godfather, Dave Cornoyer, was one of my teammates. I had met Dave at the North Attleboro YMCA after a day of Guard training, when I came in wearing my fatigues, looking for a pickup game. He saw me standing alone, came over, and started talking. We've never stopped; he's a grandpa now, and I call him, with great affection, gramps and geezer. He's six foot five and still a very good hoop player.

When we learned that the game and the christening were both set for 11 a.m., I got the christening ceremony moved up an hour. Dave and I both wore our basketball clothes under our suits. Once the ceremony was over and everyone was moving on to the party, we made some excuse and ducked out. By the time we arrived at the game, our team was down 15 points.

We both played like men possessed, brought the team back, and ended up winning in the end, and Dave and I went back to the party carrying the massive trophy, which we plunked down right next to the very pretty sugar-frosted white christening cake. I think at that point Gail knew how much I loved basketball and that it was a preordained conclusion that Ayla and any of our future children would play basketball as well.

But Dave went one better. September 12 is my birthday, and it also happened to be the date of another one of our league finals, this one set for six o'clock that night. The team was tough, it had several former Celtic draft picks, and without Dave, we were sure to lose.

Back then, next to nothing stopped us from playing. Early in the afternoon that day, Dave's very pregnant wife, Ellen, started having contractions. And Dave and I kept calling each other. At first, he said, "I think she's OK. She's going to wait until tomorrow." Everything was on for the game until about three o'clock, when Dave called to say that he had to go in to the hospital. He went into the delivery room wearing a hospital gown with his basketball stuff underneath. The birth was very quick; Emily arrived around 5:15. All of the family was very happy, the baby was in Ellen's arms, everyone was smiling, and Dave was pacing back and forth. The hospital was twenty minutes away from the gym. Finally Ellen said, "All right, go. Go." Dave gave her a kiss and bolted from the hospital. He arrived at the game with a stack of candy cigars about two minutes before start time. Our team won the game and the trophy, which I think we let him take back to his wife.

As the years went by, Dave and I continued to play basketball, and it's a tradition we've handed down to our children. Our daughters, Ayla and Emily, played together for a while on the same AAU basketball team, and both girls eventually earned college scholarships for basketball, Ayla to Boston College and Emily to Providence College.

Gail had such a difficult time recovering from Ayla's birth that we weren't sure we would have more children. We were struggling with our jobs and with her commute, and trying to manage without much help from either of our families. It seemed that we were at capacity, until we spent a night watching Ayla's baby videos and, with tears in her eyes, Gail said, "Let's have another baby." A lot of emotions went into the decision, but we decided to give Ayla a brother or a sister.

Gail got pregnant again immediately, but because she had developed blood clotting issues, she was considered high-risk and had to be constantly monitored. After a few months, it was clear that she had to take a leave of absence from her job in Hartford. I gave her the ritual three-times-daily Heparin shots and kept commuting to my practice in Boston. At the six-month mark, everything seemed fine. Then while I was at work, I got a call that her water had broken. She had nearly three months to go before our second baby was to be born. She was sent to Providence, because it was marginally closer to our house and had an OB-GYN department for high-risk births at its Women and Infants hospital. I rushed down to the hospital, driving almost one hundred miles per hour, and I remember walking through the bright, fluorescent, sterile preemie ward, looking at all the babies behind the nursery glass, babies no bigger than my hand, and being presented with a stack of bewildering legal consent forms to sign. The doctor told me that if it were a low rupture, the baby would have to be delivered, over eleven weeks early. But it was a high rupture. The doctors kept Gail in the hospital for about a week and gave her some powerful drugs to stop the contractions, and somehow the ruptured membranes resealed. She was sent home and put on bed rest, no work, nothing, but she made it to term. As with Ayla, I cut the cord when Arianna was born. We settled on the name Arianna in an almost dreamlike state—Ann was a name that our mothers shared. We also like the name Ariel. I kept playing with the sounds of the two names, and came up with Arianna.

In that moment, everything seemed perfect.

Within a few weeks, though, things seemed to be off-kilter. I could see that Gail was not herself, that she had changed. She was constantly upset with me, upset with the world. The slightest little thing would set her off. I couldn't do anything right, and one day, she took off with both girls for her sister's house. But that didn't make things any better. She appeared lost, no longer the same strong, capable woman that I had married. I watched her despair, and I thought that she might feel desperate enough to possibly hurt herself. After I talked and talked with her in the most patient way that I could, she finally recognized that things were different and that she needed help, although neither of us knew what was going on or what type of help she needed. We got her to a hospital, and in the ER, she was diagnosed with postpartum depression. I didn't know anything about it, and the obstetrician had never mentioned it, even though pregnancy complications are a major risk factor for postpartum depression. Gail had been under tremendous stress with her work, the rupture, and the forced bed rest before Arianna was born.

In retrospect, all I can say is that thank God Gail had

the intelligence and courage to say, "You know what, this isn't me," even in her horrible situation. Gail was admitted for treatment, and I stayed home by myself, with an infant and a two-year-old. As word got out, thankfully, our neighbors rallied to help however they could.

Like many new dads and husbands, I now found myself in uncharted territory. I went to see her in the hospital, and my outgoing, hardworking, no-nonsense wife was struggling to return to her old self. I prayed all the time for help and for guidance for both of us. I remember walking out of the hospital, past people in wheelchairs, their eyes glazed over from medications, and wondering, Is this what my life is going to be like? Two kids and my wife in the hospital? Will I get back the woman I married? Or is she going to be lost forever, this woman I fell in love with, the mother of my children? Is she going to be gone? Would I have to find a new and different way to try to reach the woman I love?

Even when Gail returned home a short while later, there were still trying and difficult moments, and continual ups and downs. I wanted to make sure she had every support with the kids, to try to lessen any of the burdens she had around the house and her anxieties about returning to work, and as a husband and a father,

I knew that I needed to be very, very patient. Gail's mom tried to help, along with our friends, including our dear friends Dave and Ellen, who dropped by to act as intermediaries when things seemed overwhelming. There were days that were so difficult they were ready to take bets that our marriage wouldn't make it to seven years. But Gail and I would not give up. Here, having witnessed some of the worst of divorce was both a blessing and a curse. When we married, neither of us had a good model of marriage in our own families to draw upon. Divorce to us meant not simply a breaking apart, but abandonment, uncertainty, and for our children, a loss that is impossible to quantify. It was not as if we spent our days thinking about the failures of the past, but neither of us wanted to repeat the devastation of having a broken family. We did not want that for ourselves or for our young daughters. In part, the experiences of our parents made us both stronger, as individuals and as a unit. We were determined to fight for our marriage, for our family, for the love we had for each other. We took nothing for granted, and we were determined to succeed.

Together, we had to learn how to be patient, even how to argue, how to forgive the small transgressions and the stupid stuff. After those early years, with real persistence, our lives finally settled down. Gail got better; we found good care for the girls in the neighborhood and got back into a routine. Our main disagreements were either "Why did you spend that amount of money on this frivolous stuff that we don't need or can't afford?" or "How come you are late?" It was usually silly logistical crap. And we haven't quite grown out of it yet.

Around 2004, Ayla was playing in a high school softball game. We tried to go to all of our kids' games. This one was at four in the afternoon, and since Gail worked the morning shift, we were planning on going together. The night before, I said, "Honey, why don't we leave at two forty-five, have a relaxing drive, enjoy ourselves?" I got home at 1:30. The minutes ticked by. I was ready at 2:45, when I had told her we should leave. Three o'clock came and went, 3:15, 3:30. She wasn't answering her phone. The game was at four, and I was freaking out. Finally, she came moseying up the driveway with all kinds of junk from a yard sale in the back of the car, and said, "Oh Scott, look what I got." I was now ready to blow my top. "I asked you to be home—" "I know, but they had, such great deals." And I came right back, saying, "I don't care. I wanted to go see the game. I asked you to be home. You're so inconsiderate." Any guy reading this right now will know exactly what I'm talking about. And any woman knows exactly what I'm talking about too. She's out shopping; I'm waiting. That's what we fight about. So she said, "Fine, fine. You go. I'll meet you there." Gas cost about $4 a gallon; the game was an hour away. The answer was no. I told her, "Get changed, and we'll go. Just leave the yard sale junk in the garage." "Fine," came her reply.

At this point in our lives, we lived in a cul-de-sac. There were trees in the front, and it was hard to see our house. About two minutes later, I was standing next to the car and Gail came out. Her clothes were under her arm and she was totally naked. And she said, "How's this?" with a cocky, teasing smile on her face. And I said, "It's pretty good." She answered, "You want to stay home or do you want to go to the game?" And I calmly responded, "I'd like to go, but you can still get in the car." So she did. She jumped into the car totally naked except for her silly grin and drove almost twenty minutes with me before she pulled on her dress. Just to prove a point. And she proved it. I'm driving down the highway and reevaluating whether I should have been so stuck on going to a high school softball game, even if my daughter was playing. To this day, when Gail's running late and I'm getting annoyed waiting, she'll look me right in the eye say, "You know what, you keep it up, and I'm going to get naked again." And each time, we both end up laughing until it hurts.

I also think that as we've grown as adults, we've come to see a wider perspective, to understand that all families face trials and tribulations and that the most perfect of homes on the outside may contain silent suffering within. It was only when I was long grown that I learned that the Healys, the family I had longed to join when I was ten or eleven, had not been nearly as perfect as I imagined. Looking back, there were signs. There was the time in the pool when we were playing a game of water volleyball and my side was beating Mr. Healy's, and he grabbed me after a play when I had spiked the ball on him and held me under for a little too long. When I came up for air, I saw a look of fury in his eyes that I knew all too well. Or the time in junior high when my team beat Jimmy's during basketball league. Afterward, I was walking back to our apartment in a snowstorm, the fine, icy flakes falling fast and hard. I had my thumb out to hitch a ride. Mr. Healy's car came down the road. He looked at me, hitchhiking, and accelerated the engine to drive past, leaving me in a cloud of ice because I had been on the team that had beaten his son. The Healy family would one day fracture; even the siblings would break apart. I no longer believed in a perfect home. I believed in a home that communicated and that never forgot to say, "You are my love."

Most television reporters end up crisscrossing the country to navigate their careers. Gail was deeply fortunate to be able to stay within striking distance of our home. After Arianna was born, she took another job in Western Massachusetts, but the commute and the aftereffects of postpartum depression made it challenging for her. She started looking for something closer to home. She found some work on the Boston ABC affiliate, WCVB-TV, Channel Five, doing stories for its *Chronicle Show* and cohosting a show on Lifetime Television with Dr. Penelope Leach, called *Your Baby and Child*, after Leach's famous childcare book of the same name. Then, in 1993, when Arianna was turning two, a full-time position opened up at WCVB-TV. For most of her time there, Gail had the morning shift, which meant she had to leave the house between 2 and 3 a.m. There were many mornings when it was tough on us both, but she had a great job, she was fulfilled, and we could pay our bills. The lack of sleep was particularly difficult for us, especially for Gail. Fortunately, I've been a light sleeper since high school, my battles with Larry, and my basic training days in the military. At the slightest sound from either of the girls, I would be awake.

It was now my responsibility to get the girls up,

get them breakfast, get them dressed, get them to the sitter or to school. Many days I would make dinner, do the shopping, do the laundry. I always did my own ironing too. Gail has never needed to touch one of my shirts. All those skills that my grandmother had taught me came in extremely handy. In fact, Gram had taught me everything but how to dress two girls. Because she was older and able to get her clothes on, I let Ayla dress herself, and I let Arianna pick out her own clothes. They would choose anything in their room, stripes and solids, plaids and leopard prints, and put them together, sometimes all at once, among shirts, sweaters, pants, and socks. I also had a unique strategy for fixing their hair. I'd have each one bend over and then I'd grasp the hair in one hand and wrap a scrunchie around it with the other. When Ayla or Arianna stood up, most of her hair was centered in a ponytail-type contraption on top of her head; each girl bore a very strong resemblance to Pebbles in *The Flintstones.* In the afternoons, Gail would be home to pick Ayla up when she got off the bus and then to collect Arianna from preschool, and most days, the only words Gail could think to say when she first saw each of them would be, "Oh my God." Ayla invariably looked like a tomboy ragamuffin. Arianna resembled a beautiful doll, albeit one dressed in a thousand different colors.

Both girls, Ayla especially, still say that they are terrible dressers because of me.

But I loved all those mornings with my girls and the special relationship that we developed, just us. I cooked them breakfast and, at night, I would often get into the bath with them, me in my bathing suit, and them splashing and giggling. I washed their hair and gently pulled the tangles out with a comb. We sang songs and drew patterns on the tile with our wet hands. Today, they will still call me when they are sick or when they need something, and I have been blessed to be close with them in a way that many dads miss.

And Gail also allowed me to have that. She helped to give me the gift of that relationship. Gail has always believed in me, and I have believed in her. Together, we have built the family that we, as children and teenagers, could only envy and dream of.

Gail also did one more thing that perhaps she now regrets. In 1995, when Arianna was only a preschooler, Gail encouraged me to make my first run for public office. In fact, the whole thing was really her idea.

Chapter Fourteen

RUNNING

I loved the law because of its crisp, cool logic, the way it demanded one look at a problem, understand its origins, and then find a possible solution. It is tangible and transparent. But for me, as a young lawyer in Boston, without family connections or old school ties, it was a constant challenge to build a practice. I had hooked myself up with a lawyer named John Brazilian, a good guy, a great teacher, and a tough boss. Much of his practice was real estate law; I would handle transactions for him and get a percentage of the fee, an 80–20 or 70–30 split. I also took cast-off cases from F. Lee Bailey's office, the cases his office didn't want. I handled landlord-tenant disputes, contracts, small-claims court cases, and divorces. I reasoned that this was like shooting all those endless hours on the basketball court: if I could prove myself, I could work my way up.

I built connections at the small-claims and housing courts. I could get cases handled quickly; I could get clients in and out. I learned the rent control laws and real estate. Eventually, John was able to leave for several weeks at a time, and I could run the entire real estate practice. And I think underneath, I liked the idea of helping people settle on buildings for their businesses, or families settle into their homes.

The other area where I excelled was divorce work. I understood it, both the legal side and the personal side. In the beginning, I represented mostly women trying to leave their husbands. One of my first cases was a woman who had had four prior attorneys. Her estranged husband was abusive, not just to her, but to me. He threatened to report me to the bar, and he unleashed a torrent of excuses as to why he couldn't show up in court—that he was sick, that he was dying. Finally, I had him brought in by ambulance to the courtroom and cross-examined him. Whatever he said, I came back with another question. I didn't quit. When the hearing concluded, the judge granted her a divorce, and he was finally gone from her life.

I spent a lot of time on the cases where children were involved. The parents would sit at the glossy conference table, gaze at my wedding ring, sigh, and say to me, "Oh, you don't understand." And I would tell them, "You know, with all due respect, my parents were married and divorced four times and three times each. I do know. I was that kid. I was your kid. So let me tell you what he or she is going through right now." I could say things to them that many other lawyers couldn't say and they wouldn't be offended, because I had lived it. I could write child support agreements so no college freshman would ever have to think of suing his father or mother for a basic allowance. I don't think in all my parents' divorces any of the lawyers gave more than a passing thought to the kids. But I did.

Divorce cases paid well, but they were inherently stressful. Sometimes the tensions were real, the pain of severing years or even much of a lifetime together, but some were simply stupid or driven by spite. During one of my last cases, I sat in the courtroom for five hours while the soon-to-be ex-husband and ex-wife argued over golf clubs, pots and pans, and scuba gear. The wife wouldn't let him have any of the things from their garage, and he wanted to take them, instead of simply going out to buy new ones, which he could have easily afforded. The judge finally asked both counselors to approach the bench and she said, "Are you both here for five hours to talk about pots and pans, golf clubs, and scuba gear?" And I said, "Judge, I've tried every single thing I can do. I've done everything, and we can't get past this." That case soured me so much on the field that I walked away from divorces and concentrated on real estate.

Of course, lawyering was only one of my professions. I kept modeling, through law school, through taking the bar, through marrying Gail, through the births of Ayla and Arianna. There was always work around Boston, and clients liked me because I was knowledgeable, professional, and punctual. And I had the credentials of having been a successful model in New York City. That guaranteed me a place as a big fish in the far smaller Boston pond. I appeared in ads for local stores and businesses and kept my portfolio up to date. And I was still serving in the National Guard. For years, I kept my three plans in place. If the law ever dried up, I reasoned that I could join the military full-time. If both stopped, I had the modeling. I was always thinking of how I was going to support my family, what I would do to make sure I could continue to provide for them. And all my professions intersected as well. I got legal clients from some of the guys I served with in the military; models and photographers came to me to review their contracts or handle other legal matters.

Bit by bit, too, my parents rotated back into my life. We would never be a family of shared Sunday dinners, nor would we reminisce over the last pickings of a Thanksgiving turkey. But we did try to make peace with what we had. After the girls were born, we cemented a tradition of revolving Christmases. In the morning, Gail, Ayla, Arianna, and I would have breakfast and sit around the tree, opening presents. Then we would hit the road. We usually went to my dad's first, because he was closer, had lunch, and then headed up into New Hampshire to see my mom. Many times, we ended back in Massachusetts, at dinner with some of Gail's family. For a couple of years, we tried to get everyone together at our house, but it was simply too hard; like a mesh of tectonic plates, the fault lines were too ingrained and too raw. So we kept up our nomadic holidays, starting in either Massachusetts or New Hampshire and working our way along the highways, grateful for the chance to share a Christmas as an extended family, but grateful, too, when at last, late at night, we could turn the key and hear the familiar click and release of the lock in our own front door.

I still played basketball all the time, but the leagues were getting rougher. The guys who hadn't stayed in shape were playing with their fists and elbows to win. In one matchup, I took an elbow to the eye. At first, I thought it was just a small cut, but my brow had split right down to the bone. It was possible to see the eye socket muscles and the veins. I needed thirty stitches, and as I lay on the ER table, I thought, "I'm playing to get in shape, not to get killed." Not long after the eye healed, we took a family trip to the Bahamas, and I saw an ad for the Grand Bahamas 5000, a five-kilometer road race. I thought: Hey, I ran cross-country; I'm fast; this will be easy. I entered, ran the race in the hot sun, and finished in the top ten, second place in my age group. I even won the race's raffle. When I got up the next morning, I could barely move, and I could hardly walk for the next three days. My legs were unbelievably stiff and sore—"sweet pain" I called it. Basketball wasn't nearly as challenging as I had thought. When I got home, I started running. I trained for local road races. The girls would do the kids' races and I would do the adult runs. I joined the Boston Running Club and trained after work, getting my time for a mile down to around four minutes and thirty seconds.

I was in one race, a New Year's Day four-mile road

race in Waltham—I think it was called the Hangover Classic—right behind Paul Powell, a state police trooper who happened to live in Wrentham. I spent the whole race drafting off him, banging and bumping, but never quite able to catch him and pass. He beat me by a few steps, and afterward we began talking and discovered that we had a lot in common, including the fact that we lived only one mile apart. I now had a running buddy; we started training and racing together. I've still never beaten him in a road race, but in triathlons, I'm proud to say that I can kick his very fast butt.

After I had been running in just road races, one of my buddies told me about duathlons, where competitors run and bike and then run again. I now had a new goal. I borrowed a bike and signed up for one. My first race was in Newburyport, a run-bike-run, two miles, ten miles, and two miles again. I won the run, but lots of guys passed me on the bike. I made up the ground on the second run. I won in my age group, and was one of the top five overall. I got two tickets to a local ski resort, and I won a TV in the race's raffle. I started thinking that these duathlons were a pretty good deal. I saved up and bought a bike, a Trek 1200 with an aluminum frame, not as fancy as many of the other bikes I saw, but the best I could afford. I also got a stationary trainer. At night, when Gail and the girls were asleep, I'd set my bike up in the living room and ride.

After getting better, I started entering longer races, doing better and better, and then entered national duathlon contests, including the National Duathlon Championship in Louisiana. Gail and the girls came to watch me in the swampy heat, red ants biting their feet and leaving welts between the straps of their sandals. They were miserable; but riding and running with them there to cheer me on, I was in heaven. I raced, they applauded, and I did well enough to earn a spot on the national team. I was one of twelve duathletes selected to represent the United States at the world championships.

Once I mastered duathlons, another one of my friends said I really ought to enter a triathlon. I've never been a good swimmer, but I signed up for my first triathlon event in Hyannisport, Massachusetts. I didn't have goggles for the swim, so I put on a full snorkeling mask and looked terribly out of place. Triathlons are always swim-bike-run; but the distances vary depending on the type of race. There are four categories of races. The Sprint has a one-quarter- or one-half-mile swim; a 10- to 15-mile bike ride; and, to finish, a 3- to 5-mile run. The next level is the Olympic, where everything is measured in kilometers: a 1.5-km swim, a 40-km bike ride, and a 10-km run. Then there are the half and full Ironmans. A half Ironman is a 1.2-mile swim, a 56-mile bike ride, and a 13.1-mile run. A full Ironman doubles those numbers: a 2.4-mile swim, a 112-mile bike ride, and a 26.2-mile run, the same distance as a stand-alone marathon. Ever since a nasty bike accident, the most I can manage is a half Ironman, but I enjoy the Sprint and Olympic competitions the most.

That morning on Cape Cod, I made a huge mistake. I started in the front row for the swim, and quickly plunged into the chilly swells for the quarter-mile crawl. Immediately, I was getting kicked and shoved by all the faster and more experienced swimmers who had come in behind. I floundered, swallowed water, and started going under. I could look up and see swimmers above me. If I didn't get control of this swim, I knew I was going to drown. I didn't panic; instead, I quickly struggled to the top, pushed people away, flipped over on my back, and began backstroking, skimming through the water, but nearly blind as to where I was headed and banging into other swimmers left and right. Finishing that swim was one of the hardest physical things I'd ever done in my life up to that point. I came out of the swim close to dead last in my age group, but after the bike segment, I began to catch up, passing hundreds of people, and then had a strong run. I finished the race in the top group and third for my age group. When the event was over, I was left with a feeling of total exhaustion, but also a feeling of exhilaration like nothing I had ever known.

Triathlons push you to the edge. They require an athlete to be good at three things at once. And, for me, they have a powerful ability to clear my mind. When I run, when I bike, when I move my arms in the synchronized, repetitive motion of the crawl, the rhythm of the movement gives cohesion and clarity to my thoughts. I can take a problem or a concern with me to my bike and work it through in my own mind. I think while I move, when it is just me alone, with no distractions.

I started swimming in Wrentham's Lake Archer, diving off the Ross family's dock near the center of town, surrounded by the old, lush Colonial-era green. I started sneaking in runs, the way I had sneaked in basketball games. I'd go to the store for milk and juice, but before I got the groceries, I'd run. I'd hop out of the car, do the three-mile loop at full speed, jump back in, crank the air-conditioning to dry the sweat, spritz on a little cologne, and rush home. My briefcase was now a second home for my sneakers and gym shorts, which Gail saw almost any time she opened it. When I ran, I'd wonder how I could fit in a swim. When I swam, I'd think of ways to fit in a bike ride.

I began competing in national and international triathlons and duathlons. I made the national team multiple times in my age group—competitions are based on age brackets, such as 35–39, 40–44, and 45–49; triathletes may be some of the only people who look forward to getting older so that they can move up into the next age bracket and be the youngest in their field. As a top-twelve competitor in my various age groups, I represented the United States at the world championships in places like Canada, Germany, Italy, France, and Spain, as well as the United States. The beauty of the races is that while you are competing against the other runners and swimmers, a race is also a bit like golf: you compete against the course, trying to maintain or beat your previous times on the terrain. I never wanted to leave a course or a race thinking that I hadn't done the best I could have done.

In this mix, I had never planned on running for elective office. But not long after Gail and I were married, when we first moved to Wrentham, I saw that there was an opening for a town assessor. I went down to the town hall and inquired. There was a position open because one of the board members had quit after having his tires slashed. (Real estate disputes can get pretty heated.) Since I was an attorney who focused on real estate cases, the Wrentham Board of Selectmen was happy to appoint me to fill the slot until the next election. I never had any serious conflicts because I worked very hard to be fair. When my term expired, I ran for the office, served a full term, and then retired to focus on my family and my law practice, but knowing a lot more about the inner workings of the town.

In 1995, when Arianna was nearly four and Ayla had already started school, a big school-budget-override issue came before the Wrentham Board of Selectmen. I went to a couple of cantankerous meetings and watched. The town selectmen were a kind of "old boys" network back then, and they were treating a lot of the parents who had come out to voice their views like crap. Afterward, I went up to a group of the selectmen and said it was an interesting meeting, adding that I had never seen local citizens treated so disrespectfully. And one of them replied, "If you don't like it, why don't you run? See how you like it up here." I had Arianna with me and she was starting to get fussy, so I didn't really respond, and left.

Back at our house, I started complaining to Gail about the meeting and what had happened. Finally, she said, "Stop complaining. If you're so angry, why don't you do something about it and run for office?" I looked up at Gail and decided that she was right. I ran, and I won. A couple of new selectmen also came on the board at the same time, and we all got along really well. Together, Charlie Farling, Scott Magane, Peter Preston, Mike Carroll, and I worked to solve many of the town's problems, including getting a better trash contract, a police contract, and a fire contract. In some ways, it was an excellent introduction for me to politics; it was the best of public service, helping my community, my town.

Then, in 1998, the local state representative for my district decided to run for a state senate seat. Jo Ann Sprague was a petite, white-haired woman with a penchant for red dresses, who had been a WAC in World War II. She was a Republican who believed in term limits, no more than three terms in any one position. Jo Ann called me and told me of her intentions, and added that she thought I ought to run. The rep spot would be an open seat, the best possible scenario for a new candidate. I would not be facing an incumbent. I had a good record on the Board of Selectmen, I had National Guard experience, I had my own law practice, and I was competing in triathlons, swimming, biking, and running, and often winning. I had a great family and the stability of having lived in the same town for over a decade. I was considered by many to be the classic well-rounded candidate. There was only one hitch: I was running as a Republican.

I had Republican family ties: my grandfather, who was born and raised in New Hampshire, was a Republican, as is my dad. But I came to be a Republican on my own. And it was partly driven by sports. The brutal massacre of eleven members of the Israeli Olympic team during the 1972 Munich Olympics had been seared into my mind the week before I turned thirteen. That event was bookended by the weaknesses of the Carter years, our faltering steps with Iran and the old Soviet Union, and the belief that we were somehow less than or the equivalent of our adversaries abroad. I believed in a strong military and in service, and in standing up to those who wanted to do harm. But beyond that, I had largely identified with Republicans as the party of fiscal responsibility and fiscal restraint. When you grow up with no money, you know the value of a dollar; you know the work that it takes to earn it and the choices that families have to make to spend it, and how difficult it is to save. When government asks families for their tax dollars, it is taking money that those families have earned, that they have labored over, sometimes in two or three different jobs. There is a sacred responsibility to spend that money with the utmost care. And to spend less of it. But when I looked around, especially at state politics, neither of those things was happening.

Even before I started, I knew it would be an uphill battle. Republican voters generally make up only about 13 percent of all registered voters in Massachusetts. About 51 percent are independents and the rest are Democrats. Many of the independents lean Democratic; they were usually raised in Democratic homes and have often left the party as a protest against the Democrats, rather than because they have any real affinity for Republicans. What I didn't realize when I got into the race was that, given the nature of modern Massachusetts politics, it would not be simply me against a Democratic opponent. It would be me against my opponent, and also against the Democratic State Committee; all the teachers' unions; the police, fire, and other unions; interest groups; and everyone who constituted some satellite of the Democratic machine. For years, to run as a Republican, for almost any office in Massachusetts, was to run largely if not totally alone. It was always me against the machine.

The larger irony is that Massachusetts was once the cradle of American democracy. Since the time of the first settlers landing on Plymouth Rock, local residents had gathered together in town meetings to decide the course of their communities. A noisy room of common voices, where everyone might have a say, was about as close to direct democratic rule as citizens could get, far removed from the notions of representation in Great Britain, where many people could not vote and, in Parliament, were represented by men who had no connection to their daily lives. In a Massachusetts town meeting, farmers in from the fields, millers, tanners, and soot-charred blacksmiths could argue over grazing rights, the building of a bridge, or taxes on barrels of rum. Whereas in the American South wealthy gentlemen planters often held sway, in Massachusetts there was a wider swath of voices in public life. And that was how the state thought of itself: wide, diverse, and open, even when the sound of an array of voices stopped.

Many scholars of Massachusetts politics trace the decline of small-*d* democratic Massachusetts to the profound struggles between the Yankee residents, descended by and large from English settlers, and the massive influx of Irish immigrants in the 1800s. Boston in particular, and parts of Massachusetts more generally, had, for all their revolutionary leanings, been fairly homogeneous places and fairly insular. Residents who could trace their lineage back for generations, no matter what their profession or financial position, were suddenly horrified by the sea of immigrants that had appeared in their midst. The man many now hail as an educational pioneer, Horace Mann, argued that local schools must become centralized and professionalized in large part to force these immigrants to, as he put it, "morally acclimate to our institutions," or in less polite language, to break the newcomers and remake them in a far different mold. J. Anthony Lukas, in his landmark work, *Common Ground*, about the 1970s Boston busing struggle, quoted nineteenth-century school committeeman George Emerson as saying, "Unless [the children of immigrants] are made inmates of our schools, they will become inmates of our prisons." The long-term result, Lukas noted, was a battle for control on the Boston School Committee between the indigenous Yankee Bostonians and the far more recently arrived Irish. The Irish emerged victorious and the prize was control of nearly all the appointments and nearly all the patronage slots within the public schools.

That scene has, in varying forms, replayed itself across Massachusetts politics, not between ethnicities, as in the microcosm of Boston, but between political parties, with the winners being the Democrats and the losers being the Republicans. Over the course of two hundred years, Massachusetts has gone from being one of the most representative states in the nation, a laboratory of small-*d* democracy, where the town of Wakefield would divide itself over passions surrounding the War of 1812, to becoming a state dominated by a single party.

And that domination isn't just at the ballot box. The stranglehold that the Democrats have over electoral politics has been extended across the entire political apparatus. It's not just the elected officials who are Democrats; it's all the people they've hired over the years. Every advisory board, most of the judges, the committees—all are dominated by Democrats. The vast majority of political contests are not over ideas or ideology, but over cronyism and keeping power and privilege; they are about horse-trading for favors, not over principled policies and who has the best position. A tragic result of this one-party rule is that very little is ever done in the open. Most discussions and decisions happen behind closed doors. There are few if any open debates; very little is written down; it's governing by handshake, backslap, and quid pro quo exchanges, which isn't governing at all. Whether in a town, a city, a state, or a nation, nearly complete one-party domination invites bad decisions, arrogance, excess, and oftentimes abuse of the most basic ethics and of public trust.

In this environment inside Massachusetts, once many politicians get elected, they feel entitled, as if the position should come with a bonus and a litany of special perks should be theirs. It's why one ex–House speaker faces up to twenty years in prison for allegedly taking kickbacks. It's why three House speakers have had to resign in disgrace. It's why State Senator Dianne Wilkerson was caught on surveillance footage stuffing ten $100 bills into her bra inside a fancy Boston restaurant as part of a bribery scheme. The money was given to her by an undercover FBI agent. And it's not just corruption. There are legislators with embarrassingly low attendance records for votes. They don't need to show up because they won't have any opposition when they run for reelection. They don't have to be accountable.

By the time I decided to run for the state rep slot, I knew about the Democratic political machine in theory. To confront it in fact was a very different thing.

The machine is so strong that two-thirds of all state legislators in Massachusetts run unopposed in most election cycles. There are two hundred seats total in the Massachusetts legislature; only twice in the last fourteen years (between 1996 and 2010) has the number of challengers been over one hundred. Most years, at best, sixty-four incumbents face any kind of battle for reelection. The same is largely true for the congressional delegations. It's not uncommon for congressmen and congresswomen to have had no opposition in their elections. In fact, the last time Massachusetts had a Republican U.S. senator was in 1972. Largely because of gerrymandering and because the numbers are so stacked against them, many Republicans don't even try.

Almost from the start, the Democrats were funneling money to my opponent in the race for the Ninth Norfolk district. She was picking up nearly every union endorsement, even though I had been a union member since 1982 and my days in New York. I finally managed to get the support of the state police and a couple of local police unions. I felt like an NFL quarterback, a little like Kurt Warner or Steve Grogan or Doug Flutie, with the other team's entire defensive line coming straight for me. I was running for my life, scrambling around, trying not to get sacked again.

What I think made the difference in that first state rep campaign was a question I was asked by one of the Democrats running in the primary for the seat. John Vozzella was a bit of a perennial candidate, running again and again. But this time, he struck me dead on. He said, "You're so busy. You're a selectman, a father, you're a lawyer, you're in the military, you play basketball, you're a triathlete. How are you going to represent the people of this district too? I can do a better job; I'm retiring, this will be my only responsibility."

I listened, thinking: boy, that's a good question. And I said, "Well, I'll tell you what. I have a lot of energy. If you want to get something done, you give it to a busy person, to a person who knows how to be busy and likes being busy. That person will always find the time. But more importantly, I'm going to run to everybody's house in the district, to meet each voter and show my commitment. And if I don't find your house, you shouldn't vote for me."

And that's exactly what I did. I spent all my free time running. People would see me in my shorts and I would spend eight hours a day running from house to house. The local cable commercial we ran, the T-shirts, the brochures, all of them said, "Vote for Scott Brown. He's running for you." I had about eight or nine volunteers. The other side had hundreds, but I ran to the houses while they walked. I ran so much that I tore my plantar fascia and had problems with my Achilles tendon. At lunchtime, I used to go into a local restaurant in the center of Walpole and stuff down two full turkey dinners, with gravy and mashed potatoes, because I was burning so many calories running up and down the streets. I won that race by a close margin, and then I ran again and won, and again, eventually winning easily.

As a state rep, I spent a lot of time trying to improve

conditions in my district. The Ninth Norfolk consists of lots of small towns—parts of it can even be considered semirural—and we often didn't get a lot of attention in Boston. I fought to get project money to repair roads and bridges and to create jobs. I was pro-business and wanted fiscal restraint. And I got heavily involved in public safety issues, which helped me win the respect of the fire and police unions. Much as my opponents wanted to, they couldn't portray me as a right-wing nut, which is pretty much the default position for characterizing most Republicans in Massachusetts. And I was out in the community. I loved going to meetings, to Veterans Day events, to town fairs, to senior centers, anywhere where there were a lot of people. I loved meeting the constituents and walking in parades. Some officeholders, because they have no competition, think that they can get elected and just disappear. But not me.

In 2004, my local state senator resigned to become executive director of the Human Rights Campaign in Washington, D.C. She was the first openly gay state senator in Massachusetts and had served for nearly a dozen years. When she left, her chief of staff, Angus McQuilken, was expected to be chosen as her successor in the special election. In the halls of the senate on Beacon Hill in Boston, Angus was known as the forty-first senator (there are only forty elected senators), and there were jokes about how he wanted everyone to call him "chief" because he was the chief of staff. In the race, he claimed credit for almost every piece of legislation that his boss had sponsored.

I was sitting at breakfast one morning, eating a bowl of Rice Chex with raisins, thinking about the race, when Gail said, "Listen, you really should run for this. You're qualified. You have the experience. You're a hard worker. You could do a good job for Wrentham and the whole district. The other people in the statehouse won't care about Wrentham or Plainville or Walpole; they don't even show up around here. Why don't you run?" I spooned up my cereal and I said, "You know, I think I will." Everybody else thought that I would get creamed.

To make matters worse for me, the election was the only special election in the state and the Massachusetts legislature passed a special law to move the date of the vote. Originally, like all other special elections, the state senate vote had its own date, but the senators and representatives moved it so that our election would be held on the same day as the Democratic presidential primary between John Kerry, the junior U.S. senator from Massachusetts, and Howard Dean, the former governor of Vermont. George W. Bush was running unopposed for the Republicans. The claim by the Democrats was that lumping all the votes together would save the taxpayers money, but most political observers knew better. In the minds of the politicians who passed it, changing the date of the vote nearly guaranteed they would have only five Republican state senators in the entire state senate, not six. It was being done purely for tactical advantage, because the voters most likely to show up on that presidential primary day were Democrats; there was no reason for the Republicans to come. Moving the election date was a way of keeping the political thumb down on any and all Republican candidates, particularly me, and to try to guarantee that I lost.

Added to that, the district itself was a challenge. It had been cobbled together to all but guarantee a Democratic seat, and it slithered like a narrow snake for forty miles, running through twelve separate towns, from Wayland to Attleboro. It had Wellesley College, a traditionally liberal female school; Needham, one of the most liberal cities in Massachusetts; and Norfolk, Plainville, and Wrentham, which were more moderate or conservative. This was gerrymandering at its worst, and it was hard to believe that I ever stood a chance.

I had to get 300 certified signatures from throughout the district—clear, accurate, and legible—from registered Republican or Independent voters just to get on the ballot, but I decided to get 300 individual signatures in each of the twelve towns, 3,600 signatures total. I wanted to get to know the voters. I stood in front of Walgreen's in Needham, Roche Brothers, Sudbury Farms, and a host of busy coffee shops, restaurants, and other businesses in the district. It was one of the coldest winters on record, but I stood outside in a big old blue puffy down jacket that my mom had given me one Christmas, and I collected my signatures, shook hands, and talked to people.

Angus raised huge sums of money from special interest groups, inside the state and outside the state. Every day, it seemed, he or one of his interest group supporters was dropping a new flyer beating up on me, misrepresenting my votes and my positions. We were civil to each other in the debates, but it was always clear that he thought the seat was his. A quick look at the map almost proved it. I would never win Wayland, and I also had no chance in Natick or Wellesley. Angus lived in Millis, but his mom lived in Needham, so he also claimed Needham as his hometown. I focused on the so-called southern end of the district—Wrentham, my home; Norfolk; Plainville; North Attleboro; and Attleboro—and I decided that I wouldn't concede Millis either. The goal was to do well in the lower part and try to contain the bleeding in the north. Attleboro and North Attleboro would balance out Needham, while Norfolk and Plainville were large enough to counter Natick, Wellesley, and Wayland. This meant the election would probably be decided in Wrentham.

Election Day arrived. I was on the phone from 7 a.m. until 8 p.m. that night, when the polls closed. I had a tradition that I always called all the people I knew in Wrentham on Election Day to remind them to vote and to ask each of them to call five people, just five, and remind those five too. Most usually do. I had friends and family holding up signs outside polling places, including my mom and my dad. Both of my parents had been very supportive of my state rep and now my state senate run. They had each given the maximum amount, $500, to my campaign, and had recruited their friends to help as well. That morning, they were each standing outside of key polling places, holding up Scott Brown for State Senate signs.

But at many of the polling spots, Angus's supporters, including his own mother and father, made it a point of walking up to my supporters and saying, "You know, you guys gave a good race. Our victory party is in Millis tonight." As my mother stood, proudly waving her sign for me, Angus's dad came up to her and said, "Well, my son is going to win. Why don't you come over to our victory party? We'll have a beer and break bread together." My mom was offended and gave him a few choice words back. She spent the rest of the hours waving her sign harder. And my friends, my sister Leeann, and the others out there holding signs for me were also offended. But it made them even more resolved. The polls hadn't closed yet.

By 8:30 p.m. the Boston media had all called the race for Angus. In Millis, Angus's supporters were partying and celebrating even before the polls were shut and the counting began. The victory call had been based on the early returns from Natick, Wellesley, and Needham, all places where I would do very badly. I had a friend planted in Angus's ballroom and everyone was walking around and calling him "senator." While Angus was celebrating, I was home having some chili with my wife and kids before I went over to greet my supporters. Darrell Crate, the head of the Massachusetts State Republican Party, was with me. Mitt Romney, Massachusetts's Republican governor, and his lieutenant governor, Kerry Healey, had worked hard on my behalf. They were driving out from Boston to meet me and wait for the results. My supporters had packed the nearby Luciano's at Lake Pearl in Wrentham, but they were watching the television reports and growing increasingly discouraged. I made a call to them and said, "Look, Wrentham, Norfolk, Plainville, Millis, none of these places has come in yet. You guys have given up. Don't give up." I could see the numbers. I was down by 15 percent. But then Millis came in, and I only lost there, Angus's hometown, by a mere handful of votes. Then Needham came in, and I didn't get clobbered. It was respectable, only about nine hundred votes down. Next, North Attleboro, one of my best towns, came in, and I won substantially there. Suddenly, I was down only a few percentage points in the overall vote totals—all that separated us was single digits. Then Norfolk came in, and I crushed him. I won in Plainville, and it was a draw in Franklin. Finally, Attleboro came in. Now we were neck and neck. Everything had come down to Wrentham. All of a sudden, some of the media folks began coming over to my election night spot. My supporters started saying, "Oh my God, Scott could win."

I knew Wrentham. It's a town where they either love you or hate you. The town could have hated me, but I also knew that I lived here; I had been a selectman; my kids were here; I was in the Lions Club for many years; I coached some of the local sports teams. I turned to Gail and I said, "I think I'm going to win this thing, honey." Wrentham had a lot of absentee ballots. It was the middle of winter, and people were away, so we had worked hard on the absentee ballot campaign. About 4,000 absentee ballots came back throughout the district. And I crushed Angus in Wrentham. I ended up winning the overall race by 343 votes.

Immediately, we knew there would be calls for a recount. And we were ready. Anticipating this move, we had put lawyers and poll watchers in every polling place to check for irregularities. Plus, there were new machines being used. I came right back and said Angus should save each town the money. There were no reports of irregularities or complaints in any of the polling places. The clerks had done their jobs well. Angus would not be able to make up those 343 votes.

Even into the next day, parts of the media were reporting that I had lost. Gail's own station, Channel Five, was still broadcasting that I had lost. She called up the newsroom and said, "Scott won." Her editor said, "Well, that's not what the Associated Press is reporting." And she replied, "I'm here with my husband. He won." They kept saying that it wasn't true, that all the blogs were saying that I had lost. And she said, "I'm telling you, he won." Finally, the television stations reluctantly reported that I had in fact won the race, and a few days later, Governor Mitt Romney swore me in to the Massachusetts State Senate.

Nine months later, I faced Angus and the Democratic state machine again in the regular general election. This second race was just as brutal. They threw everything at me, trying to make it seem that my race and my win had been a fluke. On Election Day, I beat him by about three thousand votes.

Because it was a special election, I had been sworn in almost immediately, and as I moved over to the senate side, we had a small reception in the house members' lounge with pizza for our guests. We invited state reps and local officials, and one of the people who came by was the town administrator for Norfolk. He told me about a house that an elderly family had donated to the state to provide a group home setting for mentally challenged adults. But instead, it was being used to house convicted sex offenders. The same private company that managed group homes for the mentally ill also had the contract to manage homes for sex offenders. These people were now living in the middle of a residential neighborhood, with children all around. And it had been done with very little notice to the town or to the neighbors. The administrator told me the story of a mother calling him up in tears because one of the men in the house was watching her and her family through binoculars in the front yard. One of my first efforts as a state senator was to get the sex offenders removed from that house. After dealing with the situation in Norfolk,

I learned a lot about the sex offender laws in Massachusetts, particularly how, in the way that they had been written, they contained far more protections for the perpetrators than for their young victims. I became a committed advocate to strengthen and revamp our state sex offender laws. In those months, my life had come full circle. If I could spare another child the fear and torment that I had known, or far worse, then every campaign, every freezing early morning that I had stood outside shaking hands would be worth it.

The most frustrating thing about being a Republican in the Massachusetts legislature was that we would constantly lose on most issues involving fiscal responsibility and good government. We'd work for hours on presentations and ways to move an idea forward, only to have Democrats say, "Oh, it's a great idea, but I can't vote for it. I'll lose my chairmanship. I'll lose a secretary. I'll lose a worker." The Democratic leaders exercised total control through chairmanships or office space, they controlled their votes through granting or withholding benefits or favors, and there was next to nothing that I or anyone else could do about it. When measures came up for a vote, the house speaker or senate president would sit on the raised dais and the legislators would crowd around in the well. They would all be clamoring for money or amendments or ways to bring home some perk for their constituents or for themselves. The scene resembled a medieval king being besieged by his serfs, with all the favor-buying and horse-trading. There was very little that was democratic about it.

Despite that, I had some great relationships with Democrats. It was hard not to. Republicans were not just a minority; we were at times a completely irrelevant minority. But we were a busy one. I had six committee assignments, while most Democratic legislators had only one, because there were so few Republicans to go around. I also got leadership pay because, again, there were so few Republicans that all of us received leadership pay. By my second term, five of us made up both the rank-and-file and the leadership on our side; there was no one else. If I wanted to get anything done, even on something like sex offenders, I had to find Democrats to work with. My entire legislative career was predicated on reaching across the aisle. But for the Democrats, even being bipartisan sometimes carried a price.

One of my best friends in the legislature was Jim Vallee, a fellow lawyer and National Guard JAG officer who represented the Tenth Norfolk district as a state rep. Jim's from Franklin, the next town over from Wrentham; he's a moderate Democrat, with two young daughters. We still serve in the same military unit together. One year, we even held a joint fund-raiser together. We invited people and said: give a check to Jim, give a check to Scott, whatever you want, whoever you support is fine—and the people who came loved it. The Republican Party in Massachusetts told me, "That's a great idea. Good for you." The Democrats were livid and tried to take away Jimmy's credentials for the state convention where they would be nominating the Democratic candidate for governor. But I've written campaign checks to Jimmy and he's written them to me, because we've said we're friends first. In local parades, we walk together; we carry one sign on the left and one on the right. We have often worked together to challenge fixed party politics and to put the emphasis on solving problems. Sometimes, we actually succeeded. But ultimately, too many issues came down to simple control, to how one party, the other party, could maintain its nearly total domination.

Being a Republican in the legislature meant I was unlikely to pass lots of legislation on my own. But that never bothered me. We have too many laws and regulations on the books as it is. Adding more doesn't often solve problems. I was able to incorporate amendments into Democratic bills, and to produce key bills on veterans' issues. One bill of mine that I did get passed was the commonsense idea of having a check-off box on state tax forms for returning veterans from Iraq and Afghanistan. Massachusetts provides these vets with a $1,000 welcome-back bonus and a host of special services, but most men and women in uniform had no idea that they were eligible. But we knew that veterans were filing tax returns. The box gave them the option of receiving information about the help and benefits they were legally entitled to, and although it seems like a small thing, it had a huge impact. Now returning men and women who have served their country can automatically receive their money.

I also worked to get more funding for a program called the Metropolitan Council for Educational Opportunity (METCO), which gives kids from inner-city Boston an opportunity to attend better schools in the suburbs. Many of these kids get up at 4 a.m. to travel to their suburban schools. They have to commute home after the end of the school day, after art or sports practice, and then do their homework. Many don't get to bed before midnight, and they do it all for the shot at a better education. I learned that METCO hadn't had a funding increase in over a decade, and the communities with METCO students in their schools had to make up the difference at a time when their own budgets were strained. For three years in a row, we got additional funding for METCO. I received three awards from METCO and became the cochair of the METCO caucus in the senate.

Sometimes, a large part of my job was to make legislation better. One such moment was the stem cell research bill in the state legislature, a bill that was very controversial. I was torn about it. I went and spoke to doctors, to clergy members, and to nuns and priests, and did a lot of research. The more I listened, the more concerned I became that if we didn't regulate some stem cell research, we would be more likely to have experiments like human cloning, and we could quickly cross an ethical line. The bill was very important for Robert Travaglini, the senate president. I told Trav that I wanted to come and see him about the bill. He thought I was coming in either to complain or to ask for something. I sat down and I told him that I wanted to support the bill, but I had some concerns about some of the ethical issues. And I said that I had a suggestion. I thought it would help if we put in an umbilical cord and placental tissue amendment to the bill, to allow for umbilical cord blood that is harvested from the discarded cord after a baby is born to be used as an alternative for research and also to establish a separate cord blood bank to preserve larger quantities of cord blood. Beyond cord blood, I thought the bill should also allow placental tissue stem cells to be used for research. I said: if we can do this, I think I'm good with it, and I think it will help you with three or four other votes as well.

We wrote up the amendment right there in his office, and then he said, "What do you want?" And I told him that I just wanted a good bill, that I wanted Massachusetts to have the best regulatory bill in the country. He said, "Let me get this straight. You're coming in here and you're offering suggestions and you'll help get me some additional votes, and you don't want anything?" I answered, "No, I don't." And he said, "Let me tell you something. This is the first time ever in my political career that there's not somebody coming in here and begging and borrowing and whining and complaining and telling me how they need me to do something for them." And then he added, "You know what—I'm not going to forget this."

Trav and I remain good friends to this day. He was a solid leader and always looked out for the people of Massachusetts.

In the 2008 election, the Democratic Party thought it could win back the Norfolk-Bristol-Middlesex district that I now held by relying on the massive pro-Obama turnout. The big money got on board behind the Democrats' handpicked candidate, a female psychologist from Needham, the largest town in the district. The Nurses Association and the Teachers Association heavily backed my opponent. The *Boston Globe* endorsed her, even though she had basically no experience and hadn't been involved in town meetings and had missed many crucial town votes. I had a 98 or 99 percent voting record in the state legislature; the votes I missed were on days when I had National Guard duty to perform. A couple of other papers endorsed her too, and I figured that the other side could be running a convicted criminal, and the reflex would be to endorse him or her over me. I ended up beating her by over 20 points at the height of the Obama wave in the election.

After that, it began to dawn on some of the Democrats that I run to win. Some of them, but not all of them.

Chapter Fifteen

"YOU'LL NEVER WIN"

There has been a sea of words used to describe Edward M. Kennedy and his hold on Massachusetts politics: titan, lion, a force of nature, American royalty, larger than life. But none really captures him or the wider array of Kennedys. I had a JFK ring as a four-year-old boy and I was moved and saddened when RFK was shot, even though my family was not Irish, and a good many of my relatives were even Republicans. One of my campaign strategists grew up in a home where a framed photo of Jack, Bobby, and Teddy hung on the wall. It hangs there still. For over half a century, one Massachusetts U.S. Senate seat had been occupied almost solely by a Kennedy brother, first Jack, then Ted. But their political legacy stretched back to the 1890s, to the days of John F. Fitzgerald, Rose Fitzgerald Kennedy's father, who was the political boss of Boston's North End, then a congressman, and then Boston's mayor. Across the city's harbor, in the late 1800s, Patrick J. Kennedy held sway as the ward boss of East Boston and as a state legislator. The Kennedys were entwined with Massachusetts like roots in soil, until it was almost impossible to separate one from the other.

Everyone knew that Ted Kennedy was ill, but the reality of his death was still hard to process. It seemed impossible to imagine him anywhere but Washington, Boston, and the Cape.

But that didn't prevent a lot of politics from being played with the U.S. Senate seat. In 2004, the Massachusetts legislature had revoked the power of the governor to appoint a replacement to fill a vacant U.S. Senate seat, with the idea that if U.S. Senator John Kerry won the presidency as a Democrat, the Republican Massachusetts governor Mitt Romney would not have the power to appoint a replacement. The law was enacted over Romney's veto, and Ted Kennedy himself made a personal request to have the new Senate succession law passed.

Then, in August 2009, seven days before he died, Kennedy made a second request. He wanted the 2004 law to be amended, basically changed back, so that the current governor, Deval Patrick, a Democrat, who was a close friend of President Barack Obama, could appoint a replacement for his office, someone who could serve until a special election was held. The compromise was that whoever was appointed as a replacement would not run in the special election. The legislature approved the measure, but I voted no. It was politics as usual on Beacon Hill. Ted Kennedy died on August 25, 2009. On August 31, Governor Deval Patrick announced that a special election to fill the Senate seat would be held on January 19, 2010. The primary to select the Democratic and Republican candidates would be held on December 8, 2009. And in the meantime, a Democratic interim U.S. senator, Paul Kirk, was heading to Washington.

What the professional politicians failed to grasp was how this one decision would begin to incite real anger in the voters. Voters, whether they are Republicans or Democrats, don't like to see blatant games being played with the election process. Just as in 2004, every maneuver was to ensure that the United

States Senate seat stayed in the hands of the Democratic Party.

But it was also the first time in twenty-five years that there had been an open U.S. Senate seat in Massachusetts.

I had a lot of people telling me that I should run. I had been one of the only Republicans to do well in 2008 against the Obama tide. And I was one of the few Republicans actually in office in Massachusetts. In 2009, the party had just 5 out of 40 state legislators, 16 out of 160 state representatives, and no Republicans in any executive office slots. But I certainly wasn't the only one who was interested. Kerry Healey, who was lieutenant governor under Mitt Romney and had run and lost against now-Governor Deval Patrick, was interested. Former U.S. Attorney Mike Sullivan was thinking about running. Curt Schilling, the retired Red Sox pitcher, was considering his own run, and so was Andy Card, former U.S. transportation secretary and President George W. Bush's long-serving chief of staff.

It was an intimidating field, but I was thinking about it. Thinking about it hard. I'm personally a strong believer in term limits for any office I would hold. I had committed myself to never serving more than four terms as a state rep or a state senator, where the terms are only two years. If I ran again for state senate, it would be my last time. The only other office I had considered running for was state auditor, because I thought I would have the background and skills to make a difference. Most of the other slots didn't really interest me. I certainly hadn't sat around plotting to run for the U.S. Senate. But, like a lot of people in Massachusetts, I was upset by the political maneuvering around the appointment of a replacement for Senator Kennedy, and I was upset by a lot of the things I saw coming out of Washington, particularly the runaway spending and the nasty, partisan politics.

This time, Gail was adamantly opposed to my running. We had over a week of back and forth, of my saying, "I want to run," and her saying, "No, you don't." Finally, she threw up her hands and said, "If you want to make a fool of yourself, go ahead."

I would never have made it to the point of even considering a run for the U.S. Senate without Gail. For over twenty-three years, she has been my touchstone and my bedrock. She encouraged me and signed on to my political runs, even knowing that they could build a torturous divide between us and would subject the family to the blood sport of Massachusetts politics. Gail is very dedicated to her profession as a journalist and newswoman, a profession that requires her to be completely impartial and fair. When I entered public office, there were entire areas of my life and her life that had to be walled off from each other to maintain that independence, impartiality, and professionalism. There were, quite simply, things that we could no longer talk about, experiences that we could no longer share. When I was criticized or maligned, she had to stay silent, and she did. She could not campaign with me, and she had to stay home on election night. When she wanted to be there, she couldn't. And she did it because she had made a commitment to her profession and a commitment to working on behalf of our household, so that we could give our daughters the best education, allow them to play sports and pursue their own dreams. She stayed as the silent wife at all costs, and I am grateful and humbled by how hard it was for her to do that.

Now, at the end of August, even Gail thought like everyone else: this was the Kennedy seat; it belonged to the Democrats; why should I run and lose 70–30 in a special election? "They've got strong Democrats. They've got multimillionaires running, the state attorney general, and a congressman. You can't win. You just can't win." It was my fiftieth birthday, and we had a trip planned, back to Aruba, where we had honeymooned twenty-three years before. She was set on this trip. But I was thinking, "I can win this race." I learned early in life that no self-respecting basketball player ever leaves the court before taking his best shot. This was no different.

I met with a team of campaign experts: Eric Fehrnstrom, Mitt Romney's former presidential campaign spokesman; Beth Myers, who had been Mitt Romney's chief of staff; and Peter Flaherty, a former prosecutor. We started to explore what it would take for me to get into the race. Early on, Kerry Healey decided not to run. I called Mike Sullivan myself to ask him whether he was going to run. Then, on Labor Day weekend, I started calling all the state senators and representatives. I told them I was going to run and asked them for their support. A lot of the people said that they would support me, unless Andy Card ran. Then they would support Andy.

In many ways, Andy was the logical choice, especially after Ted Kennedy. He knew everyone, not just in Washington, but around the globe. Kings and queens and prime ministers all knew Andy. He had a great work ethic, a great demeanor, and a strong family, and I'd always admired him. He was a known name, and someone who could probably raise a lot of money, but he also hadn't lived in Massachusetts for any length of time in nearly two decades, and he was closely identified with George W. Bush, who, eight months out of office, was not even close to being well liked in the state, to put it mildly. He would have some work cut out for him.

But no one knew if Andy was going to run.

I had put out all kinds of feelers trying to reach Andy, but I had never met him personally and he probably didn't know me from Adam. But I did know his brother-in-law, Ron Kaufman, a Republican National Committee man, who had campaigned with me back in 2004 when I was trying to win the state senate seat in a special election.

I was in Boston on state senate work when I got a call from Ron. He said, "Can we meet? Andy's in town and he wants to meet." Ron and his driver picked me up outside the statehouse in a dark, unmarked SUV with tinted windows, and we drove the short distance to his home in Beacon Hill. When we arrived, Ron said, "We wanted you to meet because Andy's thinking very strongly about running." We began with the usual pleasantries and my telling him what a great guy I think he is. And then Andy said, "I'm really thinking about running." He was planning on announcing within a couple of days; I was planning on announcing that very night at the Massachusetts Republican State Committee meeting, where all the state party activists would be gathering. Andy said, "You are? I'm going to be there. I'm going to speak."

And Andy told me why he wanted to run, and that he was planning on doing it.

I told him, "Andy, I have great respect for you, but I've been trying to reach you for the better part of a month. I've had people trying to reach you to see where your head is at, and we haven't heard anything. I'm running out of time." Like every other potential candidate, including Andy, I needed to get ten thousand certified signatures to get on the ballot. And then I told him, "Andy, I'll beat you."

That upped the testosterone level in the room. He replied, "Well, I don't need to be threatened by you, you know." I said, "I'm not threatening you. I'm just telling you honestly that I would beat you because I have a team ready, I've been working in state politics for twelve years, I've got four hundred supporters throughout the state ready to go tomorrow. I can raise money, and I'm on TV and radio regularly, sometimes every day. People don't remember you. They don't remember who you are. They remember that you're the guy who whispered in the president's ear on 9/11. And they have great love and respect for you for that, but they also know that you were very close to President Bush and right now President Bush isn't very popular and you will be tied to that. And I don't have that problem. And I think you would lose and I would beat you."

You could have cut the tension between the two of us with a knife. Andy Card was offended that I would call him out like that, the first time that we met. And I probably was being a little disrespectful; I was fairly pissed, because I had been working to line everything up and we were ready to go. But he was also pissed. Now, Andy started talking.

I listened to him intently, and we kept talking and talking, and I could see the passion he had about wanting to be in this race, about wanting to do it and to do it really well, and how important it was to him, to his family, and how much he cared for his country. I knew he was serious and would do well if he won. His passion truly moved me, and at that moment, I had an epiphany. I said to myself: This guy really wants to be a United States senator. I've always liked him, and I would love for him to be my United States senator. He'd really give them a battle. And who am I? I'm just Scott Brown from Wrentham. This guy's been the chief of staff to the president of the United States. Ten or fifteen minutes had passed, and I decided that I would throw a curveball, and mean it. I decided that I would support him. I said, "You know what, Andy, I really think you want this." And he said, "I do." And I said, "Well, I'll tell you what I'm going to do. I will wait until Friday at twelve o'clock noon. And if you tell me that everything's in order and you want to do this, then I will be with you. I'll support you tonight at the meeting. I will tell people that you are the best person for the job. I'll say that I'm supporting Andy, and I'll ask everybody else to do that. And I'll do it tonight."

It was a curveball all right. Andy looked at me in disbelief. He practically asked, "Are you kidding me?" I've never seen a look quite like the one he gave me; he was completely floored. And he said, "You'd do that for me?" I answered, "Absolutely."

Andy shook his head slightly and said—I still remember his exact words—"You're an amazing guy. Everything people have said about you is absolutely right. I can't believe you'd do that for me." I told him, "Look, this is about governing, and because of your experience and who you are—you're one of my heroes too—you'd be a great senator and I think you're a great guy. I mean this isn't about personalities. When I'm telling you this stuff, I'm just speaking very frankly. I'm not trying to be a jerk or hurt your feelings. I'm just being honest with you. Even though I think I can beat you, I'm going to drop out and support you." And that was it. I went back to my state senate office and told my chief of staff, Greg Casey, that I was going to support Andy Card.

That night, the Republican State Committee meeting was packed with people and cameras, a rarity for this event, which is hardly ever covered. Everyone was expecting a showdown between Andy Card and Scott Brown. I got lots of questions about what I was going to do, and I didn't say a thing, and Andy didn't say a thing. I got up first to speak. I gave what I thought was a good speech, kind of like a pep talk, which led up to the line, "And I've been working hard to bring this party together and this position is important, it's a real opportunity for us, and we need to come together as a party and that's why I'm," and I paused, and said, "endorsing Andy Card as the next United States senator from Massachusetts. He has my full support, and I encourage each and every one of you to do the same." The crowd was silent, like a delay in transmission, and then suddenly everyone got up, cheered, clapped, and gave a standing ovation. Andy himself was standing right behind me as I gave the speech. I don't know if he really thought I would do it, but I did, and I turned and gave him a handshake and a quick hug, and wished him well. He stepped to the lectern, gave a wonderful speech, and that was it. Some of the people there told me that they couldn't believe I had done it, and asked why. I answered, "Because it's the right thing. It's going to unite the party and make us stronger." And in retrospect, it did. It brought the party together, and it wasn't about me, or any one person; it was about a strong candidate and a good race. It was about winning the seat. It got people excited.

I waited all week. My group of campaign consultants, Eric, Beth, and Peter, went ballistic over what I had done. I hadn't told them before I stood up to give my speech. But my mind was made up. I felt comfortable that this had been the right decision. I heard that Andy had ordered ten thousand signature pages to get on the ballot, that he was going out and doing different things. I thought it was over. I paid for the trip to Aruba for Gail and me, and I reserved my spot on a trip that a bunch of my buddies and I were planning on taking to Las Vegas to celebrate our fiftieth birthdays. I made my plans to get on with the rest of my life.

At last, Friday arrived: Friday, September 11. And I didn't hear from Andy. I placed a couple of calls that afternoon, and nothing. A baseball game was on, and I sat down to watch it and have a couple of beers. Around 9 p.m., my phone rang. It was Andy. He had decided not to run. Now it was my turn to be completely shocked.

I started working the phones; I called my team; I called my chief of staff. By 2 p.m. the next day, I had a hotel reception room rented and campaign signs printed, I had my speech ready, and I had people there. I got up in front and announced my candidacy for U.S. Senate. My case was simple: I believe higher taxes will further weaken our economy and put even more people out of work, while in Washington, politicians mistakenly believe that spending more money and increasing the size of the government is the answer.

I was very blunt, saying, "They are wrong." It's the private sector, small businesses and entrepreneurs, which will get the economy going again. Government can help sometimes, but it is also vital for government to know when to step out of the way.

I promised that I wouldn't take my orders from special interests or from Washington politicians. In the state legislature, I had never taken my orders from the entrenched Beacon Hill establishment. I was not part of the insider club. I promised that if elected, "I will be guided by what's right for the people of our state. On every issue, I will ask myself: Am I representing the people of Massachusetts? Will this issue empower them, or benefit only big government? Will it raise taxes or increase the federal deficit? Will it protect or create jobs? I do not want to go to Washington to serve the interests of government. I want to serve the interests of the people of Massachusetts."

I also took a firm stand against the one-party rule that had gripped Massachusetts, saying that we have eleven elected officials from the majority party in Washington, as well as a special Washington office to represent the state's governor. All of these officials usually vote the same way and often take their orders from the same special interest groups and political leaders. I asked, "Does Massachusetts need another elected official to merely rubber-stamp the policies of one party or one administration?"

I stated that I believed Massachusetts needs someone who is an independent thinker and an independent voter. As a legislator, I have always believed in good government. I've always worked across party lines to ensure that when there is debate, the debate is factual, spirited, and never personal, and that the interests of the people of this state are always paramount. My final lines were, "Already, my opponents have started pandering to the special interests, promising to support their pet projects. That's not the way I operate. Because I don't owe anybody anything, I'm free to tell the truth and fight for what's right for all of the people of Massachusetts, no matter their political party. That is the type of senator I will be—free to speak my mind, and act in the best interests of the people I represent."

That was my pledge. I never deviated from it. The

day I made that pledge happened to be my fiftieth birthday.

Gail wasn't the only one who thought I couldn't win. My consultants didn't believe it either. "If you run a credible statewide campaign this time, you'll put yourself in a good position to run for another statewide office," they told me. They thought that the U.S. Senate run would be a good way to position myself for lieutenant governor or for attorney general, because I was a lawyer and a JAG. A lot of the media and radio hosts were saying the same thing, speculating about what I would run for next.

I never bought into that thinking. I said, "With all due respect, I'm going to win this thing. If I'm going to run, I'm running to win. I'm not running as a steppingstone for something else." The thought of running one massive statewide race only to turn around and do it all over again in another nine months made me sick to my stomach.

The consultants nodded their heads and privately concluded that in the next election cycle, the attorney general slot would be open because the current AG, Martha Coakley, was running for the U.S. Senate seat as a Democrat. All the smart, conventional thinking was that Martha Coakley would be the next United

States senator. And that was exactly what Andy Card, Kerry Healey, and everyone else had been hearing. They got out of the race, and I got in. To paraphrase Robert Frost, it was the decision that made all the difference.

I signed a $50,000 campaign contract and I remember lying in bed with Gail that night and saying that I hoped we could raise enough money to cover the cost of the contract. Gail was lying next to me saying, "Oh my God, are you sure? I don't know if we can afford it."

I just wanted to be able to pay the bills. I didn't want to owe anybody any money from this run. I remember some days praying, "Dear God, please let me be able to pay the bills. That's all I ask. I don't want to owe anybody any money. I don't want to read in the paper that Scott Brown lost and people couldn't get paid."

In my mind, I spun out all the financial calculations. I thought I could be competitive if I was able to raise just $700,000. If I had $1 million, I decided I could be really competitive. With $1.5 million, I thought I could win it. With $2 million, I knew I could win it. But right now I was worried about that first $50,000. I was not sure how I would do it, but I went forward.

Hours after I announced in Boston, I started trying to get signatures to get on the ballot. Early that evening,

I was heading to friends' houses, and I was also trying to reach Curt Schilling, the baseball great, thinking maybe I could talk to him and eventually get his endorsement. My friends weren't around; Schilling was nowhere to be found; then I got a call from Greg Casey saying that Schilling was just finishing up a charity event at the Eagle Brook Saloon in Norfolk. If I hurried, I could catch him. I raced over, and there was no Curt Schilling, but Gail, being Gail, had gotten almost one hundred of my friends together for a surprise birthday party on my actual birthday, even though I had just announced for the U.S. Senate. Everyone I had been trying to find to sign my signature sheets was there. And my friends were all really excited that I was taking the shot. They told me that they thought I could win this thing, that I would do a good job, that they knew me, and that they knew I would work until I dropped. And that's what I did; I worked until I dropped.

That night I started getting signatures and enlisting volunteers. It was a great end to my birthday—and probably the first time Gail has managed to keep something a complete surprise.

When it looked like a showdown between Andy Card and me, the media had been fascinated. Phones were ringing off the hook in my office, and reporters camped out in front of my house, so I would have to go out through the back door. When I announced my candidacy, it was pretty crazy too, but within a few days, the frenzy died down because no one thought I could win. My one primary opponent was a perennial candidate, Jack E. Robinson III, an African-American Republican and a millionaire, who had run against Ted Kennedy in 2000, and had gotten about twenty-five thousand more votes than the Libertarian candidate. Kennedy had beaten them both by over 70 percent.

This time, the Democrats were getting all the coverage. It was Coakley against a very capable congressman, Michael Capuano; Alan Khazei, a civic activist; and multimillionaire Stephen Pagliuca, a part owner of the Boston Celtics, all good candidates with different strengths. While the Democrats were running against each other, I was out hustling votes, going everywhere I could. I went to senior centers, to town meetings, I knocked on doors, made phone calls. I went to any big event where I could meet people. I stood outside Red Sox and Bruins and Patriots games. I went to county fairs, including the Topsfield Fair, where I had gone all those years as a boy. I made my race into a retail politics race. I wanted to shake as many hands as possible, to meet as many voters as possible, to get out there so people could know my name. I didn't care whether they were Democrats, Republicans, or Independents.

I just wanted to reach voters. In the middle of January, it would probably be a low-turnout race. I calculated to myself that six hundred thousand votes might be enough to win. But there was something else that mattered about going out and shaking hands and listening to people. People were energized. They were engaged; they were angry. And they did not like some of the things they saw coming out of Washington.

But the Boston and Massachusetts media largely thought the Republican race was a joke, and their lack of respect showed daily in their lack of coverage. The Republican primary had a near-total news blackout. When I did a debate with Jack, the room was barely filled.

Meanwhile, my team was trying to raise money. From the afternoon of my announcement, we were fortunate to have on board a great campaign manager, Beth Lindstrom, who had worked as director of consumer affairs in the Massachusetts State House and had been an official in the Romney administration. She brought a tremendous amount of credibility to the race and was extremely capable. Beth's job was to keep the office together, hire people, and manage our meager budget. We were lucky, very lucky, to pull in $10,000 a week. I kept telling my political team that I was going to win the race. And the team kept telling me, "Yeah,

yeah, you just get ready for the next run." I told them I wasn't going to run again in another nine months. The race I was running to win was this one.

We put up a Web site, had a blog, wrote letters to the editor, and got on Twitter, Facebook, everything. We did direct mail to all the Republican State Committee people. Mitt Romney had endorsed me early on, and his political action committee wrote the first big check to my campaign. I started going to town hall meetings in different areas of the state and continued calling in to talk radio shows. I knew almost all the radio hosts and a lot of the print and TV people because while I was in the state senate, I was one of the spokesmen for the Republicans and would be called upon to discuss various issues. Especially with the talk radio hosts, I would be one of two or three people tasked with going out and answering their questions and speaking about the issues, so we already had good and positive relationships. I'd be driving around in my truck, and I'd call in on my cell phone, and they'd put me on the air. My team of consultants wasn't that happy that I was doing so much unscripted talk radio, where anything could happen. To me, that didn't matter. I enjoyed answering the questions honestly without all the handlers who followed the other candidates around. My attitude was: Why do I need handlers? Either I know the answer or

I don't. Meanwhile, all the political pros were basically writing us off.

On November 16, three weeks before the primary vote, I headed down to Washington with Peter Flaherty and my campaign policy coordinator, Risa Kaplan. We had four appointments: a 10 a.m. meeting with former UN Ambassador John Bolton, a noon one-on-one meeting with Senator John McCain, a meeting from 1 to 3 p.m. at the Heritage Foundation's think tank for foreign policy briefings, and a 3:30 p.m. meeting with the Republican Senatorial Committee. We had already learned that the Massachusetts special election wasn't even up on the senatorial committee's Web site, so we were prepared for that meeting to be a waste.

The McCain meeting was great. John McCain endorsed me and wrote a check to my campaign, which, given our struggles to raise cash, was a huge boost. In fact, it was all great until 3:30, when we got to the Republican Senatorial Committee offices. One staff member was there to meet with us. I made my five-minute presentation about how I could win or at least make it very close, and how this race could send an important national message. I knew it was probably the only chance I would ever have. I talked about how the Democrats were energized and engaged, and the Republicans needed to be as well. The staffer sat, nodding as I spoke, and ended with the recycled lines, "Keep up the good work, Scott. We need more people like you. But we don't take sides in primaries," although of course they do.

"I need your help for the general election," I told them. "It's going to be a short race, just six weeks."

"We'll get back in touch after the primary," the staffer said. "We wish you the best of luck, and we'll be in touch." He might have saved himself two minutes if he had just said, "Don't let the door hit you on the way out."

That night, I was invited to a Republican Senatorial Committee reception at the Grand Hyatt in Washington. It was for members of Congress, political figures, donors, and hopeful candidates. Andy Card met us at the door and generously took me under his wing. He introduced me to all the heavy hitters in the room, to Senator John Cornyn of Texas, head of the senatorial committee, and to Senator Orrin Hatch. I walked around meeting people I had only seen on television. California Republican Carly Fiorina was there, as were all kinds of high-profile candidates. Andy prevailed on John Cornyn to mention me briefly during his remarks, and I made sure that I shook hands with nearly every person in that room. Most people didn't even know there was a race in Massachusetts. It was almost embarrassing, and we couldn't get anyone to commit. I felt bad for Andy as well; I hoped he wasn't wasting his own political capital on me.

Not only did the national politicos not realize that there was a race in Massachusetts; neither did some in the national media. Not long after my trip to D.C., I was driving to an event in Western Massachusetts, and flipping around the radio dial. I stopped on the *Laura Ingraham Show*, a conservative talk show, which that day had a guest host. He was talking about how in November 2010, we would be taking things back. "We'll make a difference next November." I was sitting in the car saying, "November! We have a race on January 19, 2010. This guy is clueless." He kept talking about how nothing could happen until next November, and I was getting livid. I dialed my cell phone, and somehow I got through to the show. I told the producer, "Hi, this is Republican State Senator Scott Brown, and I'm running for the United States Senate." He said, "Where?" And I said, "In Massachusetts."

"In Massachusetts?" he said. "What's going on in Massachusetts?"

"There's a special election in a couple of months," I told him.

"Hold on," he said.

He probably did a quick Google search and then

he came back on the line and said, "Tell us about the race." I answered that I'm a Republican, I've been a town assessor, a selectman, a state rep, and a state senator. I've spent thirty years in the military. And then I finally just asked, "Do you guys even know there's a race? Because I'm sick and tired of listening to you talk about November 2010. There are other things happening before then." During a commercial break, the producer put me on with the substitute host himself, who began by asking me, "So, is Mrs. Kennedy down there voting now?"

I replied, "Mrs. Kennedy? What do you mean, Mrs. Kennedy? There is an interim senator now. They changed the Senate succession laws in Massachusetts so they could put in an interim senator, Paul Kirk, to push the president's health-care bill and other parts of his agenda through." Then I told him, "Let me get this straight. This is Laura Ingraham's show, right? And you're telling me that, number one, you don't know there's a race in two months to fill the seat left by the late Senator Ted Kennedy, and number two, you think that Mrs. Kennedy is actually down there voting for him?"

The guest host paused, and then he said, "Let me get you on the air." I got five minutes. I was able to tell the story of the election and to work in my Web site, www.brownforussenate.com. The host ended the segment by saying, "Listen, folks, if you're interested, you can go to www.brownforussenate.com. If you like what you see, then jump aboard, because we can really send a message right now."

That day, our Web site and our campaign raised $12,000.

Now I had a new mission. I wanted my campaign team to start reaching out to the national media. The Democrats had national media, national committees, national unions, and all I had were my buddies from high school and some others who were gluttons for punishment. Aside from a very fair and thorough profile in late November in the *Boston Globe* by Brian Mooney—a reporter for whom I have a lot of respect, who for years had covered politics, and who had a no-nonsense reputation as a guy who did his homework—and a few other scattered pieces, there was very little focus on me or Jack Robinson. No one was paying attention to my side of the race. The Democrats had taken up all the oxygen in the room.

Eric Fehrnstrom, Beth Myers, and Peter Flaherty on my campaign team told me it would all change after the December 9 primary. Then, they said, the coverage will have to be equal. When they write about Martha Coakley, they'll have to write about Scott

Brown. "The klieg lights will burn brightly on Scott Brown after the primary," Eric said. "There will be equal coverage starting tomorrow," Peter promised, as we—Eric and Peter, me, Ayla, Gail, Gail's sister, and the teleprompter operator—waited in a small suite in the Newton, Massachusetts, Marriott hotel on primary night. As I headed down to the hotel's basement to deliver my victory speech, Peter added, "After tonight, it will no longer be just about the Democrats." I had won the Republican primary vote, 89 percent to 11 percent.

The next morning, I woke up to a giant *Boston Herald* headline that said, "She's the One." I barely rated a mention anywhere on the front page.

Two days after the primary, on December 11, I held a news conference to question Martha Coakley's positions on spending and her economic proposals. One reporter came and wrote about how he was the only guy who showed up. The lack of coverage made me determined to work even harder. I would take my case directly to the voters, and screw the media.

After the election, the tagline before my name in many articles was either "long-shot candidate" or "little-known state senator." As local Boston political commentator Jules Crittenden said, "The national GOP isn't interested in even making a good showing for Ted Kennedy's seat" and "The state GOP is a joke." He called it all but a foregone conclusion that Coakley would win, unless she made remarkable gaffes, and I also made an extraordinary and deft effort, or there were external events, like a "Democratic health-care debacle." On Gail's station, Channel Five, political consultant Mary Ann Marsh was constantly hammering me, often launching into personal criticisms that became increasingly unprofessional.

"The writing is on the wall" was what nearly every mainstream commentator said. It didn't matter that Virginia had elected a Republican governor, or the historically blue state of New Jersey had just elected a Republican governor as well. Statewide polling had Martha Coakley as the most popular Democrat in the primary race. A *Boston Globe* poll had 71 percent of likely Democratic primary voters saying that they viewed her favorably, and of all the candidates, she was the person with whom they would most like to have a beer. Those types of numbers might put her ahead of Senator John Kerry and Governor Deval Patrick in popularity. She had just won her own primary by over 20 points. Whatever political office she set her eyes on was considered to be hers for the asking.

But I was taking a different poll. I was out shaking hands, and people liked what they heard. When

I'd ask people if I could count on their vote, they'd say, "Yeah. I don't think it will matter, but yeah, you've got my vote." But gradually, as the weeks went by, people stopped staying, "I don't think it will matter." Because by then, it did.

Chapter Sixteen

"IT'S THE PEOPLE'S SEAT"

It wasn't hard to figure out Martha Coakley's campaign strategy. It was simply to behave as if she had already been elected. She spent December networking with Democratic officials in Massachusetts and Washington and traveling around to the swearing-in ceremonies for local Democratic mayors across the state. Martha is a nice lady, and she devoted herself to the race. The media myth has become that she didn't work hard, because that's the easiest way to place all the blame for the outcome at her feet.

But unlike Martha and unlike probably most of the Republicans who had thought about running for the U.S. Senate seat, I had thought from the start that she was vulnerable. I knew her races; she had never been forced to run a tough campaign. Her elections had always been easy, while mine had always been bloodbaths. Easy is not necessarily good. It can make you complacent. Lots of people in Massachusetts politics were afraid of Martha Coakley. But I wasn't, even though I knew just how much machine support she had behind her. In addition to the state and national Democratic Party apparatus, she got the endorsement of most of the major newspapers and all of the key unions. When I drove home each night on the expressway, I would pass the International Brotherhood of Electrical Workers headquarters with its big electronic billboard. Every day, there would be an image of Martha Coakley made out of hundreds of tiny lights alongside an illuminated flag. Whenever I felt tired or worn-out, I'd look at her electronic picture, and I would think: I'm going to do one more event. I'm going to work harder. I'm going to make another one hundred calls.

I had gone to make presentations before the key unions around the state—many times Martha hadn't even shown up to speak to them, but the union officials still endorsed her anyway or largely sat out the race.

My positions were increasingly resonating with the union rank and file, with people like me, but that didn't matter to the union heads. The problem was simply a letter, the letter *R* after my name.

Throughout December, I went out to meet voters. I stood outside Fenway Park in the freezing cold to shake hands with fans before the Winter Classic hockey game, featuring the Boston Bruins against the Philadelphia Flyers. I stood outside the TD Garden to meet people going in to Celtics games. I dropped by bars in South Boston, in Dorchester, and throughout the state, and went from booth to booth, meeting people and listening to what they had to say. I'd pop into senior centers. I campaigned in front of coffee shops and hardware stores, and when I did, I always went in and bought something. I was hungry, so food was a natural, but there were also things that I needed to make repairs around the house. I figured that if shop owners were kind enough to allow me on their space out front, the least I could do was to patronize their stores.

I kept doing the radio. Each host had a very different personality, and I liked the variety and interplay and how seriously each one took his or her job and responsibilities. Howie Carr was good at mixing serious issues with a sense of humor. I knew Michael Graham from his work on behalf of veterans and their issues, and he always made good points. Jay Severin had his finger firmly on the conservative pulse in Massachusetts, and always expected honest answers to his tough questions. I enjoyed Michele McPhee, who, like me, was from Wakefield, and who focused on fighting for the little guy and law enforcement. Jim Braude, a liberal host, liked to go for the jugular, particularly with Republicans, but I enjoyed his "out there" humor and have always appreciated his preparation and fair treatment of me. On the morning drive, Tom Finneran and Todd Feinberg focused on highlighting the important political issues of the day. And I got great and vital radio attention outside of Boston. In Worcester, Massachusetts, former U.S. Representative Peter Blute, a Republican who had represented the state's third district, talked with me regularly over doughnuts on his lively morning show. Two other Worcester hosts, Jim Polito, a former chief investigative reporter for the ABC affiliate in Springfield, Massachusetts, and Jordan Levy, both at WTAG, also had me on and gave me a fair hearing. I called in to Bo Sullivan and Brad Shepard's morning show out in Springfield and spoke regularly with Ed Lambert at WXTK down on the Cape. In Lowell, WCAP gave me a great forum, and so did WXBR in Brockton. In large measure, radio and the radio hosts who knew the issues and could discuss them in-depth were what made the campaign. Apart from radio, I would appear on whatever media would have me, and I talked up the Web site, www.brownforussenate.com.

Money was now coming in at a better clip; we were able to pay our bills. But every day was still a challenge. Beth Lindstrom, my campaign manager, who had come out of political retirement to help, was working almost round-the-clock. She tried to keep everyone focused on the ultimate goal, making tough decisions and managing a small and not always experienced staff while still keeping a smile on her face. She put in countless late nights over those few months when I know she would have rather been home with her family, and I remain grateful for her dedication.

Our chief goal was to save all our money for TV ads, because if a candidate isn't on TV, that candidate isn't considered serious. During the primary, we had spent only about $40,000 on media, primarily on radio, compared with the hundreds of thousands of dollars that Jack E. Robinson III had spent on TV, radio, and mailers. I wanted every penny accounted for. We planned on buying $500,000 worth of airtime in the last two weeks before Election Day.

Where the campaign was taking off was over the Internet and the radio. We had already launched Web ads, including one before the Democratic primary. We called it our Halloween ad, officially "the Democrat House of Horrors," and it gave all the primary candidates roles in an old-style Hollywood horror film. Representative Mike Capuano was "Cap and Trade" Capuano, and there were "Second Stimulus" Khazei and "Public Option" Pagliuca. But the scariest line of all was reserved for Martha Coakley, who in a late-October television interview, when asked about her foreign policy and national security experience and whether she would face a learning curve in the Senate, had responded first by saying, "My sister lives overseas." She went on to explain that her sister had lived in London and now the Middle East, and she discussed what her sister told her about dissatisfaction with George Bush's policies abroad. But calling one's sister for foreign policy advice wasn't a great response. We had tried to use that ad to get some interest in the race down in Washington, to persuade the political operatives and the money people that Coakley was not invincible. But they still weren't interested in Massachusetts. The early polling had showed that I was down 31 points to Coakley. The race, as of mid-December, was still supposed to be a blowout.

But it didn't feel that way when I went out and shook hands and spoke to people. I was spending my day with actual voters. I'd leave the house at 6 a.m. and come back around 10 or 11 p.m. In between, what I heard was that people were scared, worried about the economy, worried about their jobs, worried about how they were going to afford new taxes in one of the highest-tax states in the nation. And it didn't feel like a blowout when I talked to radio hosts. I came back from my time on the campaign trail and told my staff, "I think we're only down by ten to twelve points." We had less than one and a half months to close that gap.

On December 17, Peter Flaherty, one of my campaign advisers, was in Washington, D.C. He stopped by the Republican Senatorial Committee for all of five minutes and told the people there that the race was tightening. They asked if there was any polling. He said not much, but it was clear that the race was tightening. I was collecting some big endorsements, starting to get attention. Once again, the Republican Senatorial Committee gave him the brush-off. Come back when you have some real numbers, they said; for now, we'll be monitoring the race.

I very much wanted to debate Martha Coakley, but she didn't want to debate me one-on-one. When she ran for attorney general, she had refused to hold a single debate with her Republican challenger, even though he was fully qualified. This time, she was even less enthusiastic. She would not agree to a live televised debate with just me. My campaign kept offering to debate her, anytime, anyplace. Her counterproposal was to have Joseph P. Kennedy—no relation to Ted Kennedy's family—who was running for the office as a Libertarian candidate, onstage and part of the debate too. She insisted upon it. We knew her strategy: she wanted people to see that he was not a member of the Kennedy family, and she wanted to dilute my presence and strengthen hers. I said fine, I'd have a debate with whoever was on the stage.

Our first group appearance was on Dan Rea's radio show, *Nightside*, to air live at 8 p.m. on December 21. It was originally designed to be a forum, and I finally got fed up with her talking around the issues, talking in half-truths, and waffling. I respectfully started going right after her, questioning her and her policies and half answers. I wanted listeners to know that I was playing for keeps. When the forum was over, she said, "I thought this was a discussion, a forum. I didn't realize we were debating here." And I replied, "Well, Martha, I'd love to debate you. You won't debate me. So I wanted to ask questions and hope that you'll provide the answers." After our forum, when she was asked why she wasn't debating me, her answer invariably was, "I already debated Scott. I already debated him on *Nightside*."

Right after *Nightside* aired, three things happened. The Republican Senatorial Committee called us, we decided on a big media buy, and in a plane flying from Amsterdam to Detroit, a Nigerian man attempted to detonate a bomb in his underwear.

On December 23, the Republican Senatorial Committee called my campaign advisers. The committee said it had a poll showing me just 13 points behind Martha Coakley. At first, Eric and Peter thought the committee had misread the polls. The numbers usually showed me trailing by 31 points. But this was a real poll, and what's more, among voters who said they were intensely following the race, only single digits separated me from Martha Coakley. The committee was offering us phone banks and technical support, and the separate Republican National Committee would kick in $20,000 cash for the balance of the race. We gladly took it. The Republican Senatorial Committee also gave us support and guidance with research and background issue reports, and I was grateful. It was one of the most valuable pieces of assistance I received at that point in the campaign.

For Christmas Day, I took a break from campaigning. Arianna and I went to a homeless shelter to serve food to those less fortunate, which was a wonderful way to give back on that day. Afterward, we did what we always do: we drove to Gail's mom's house, to my mom's house, and to my dad's, three separate Christmases with three separate families. By the end of the day, I was completely exhausted. In the middle came word that Umar Farouk Abdulmutallab had tried to detonate plastic explosives hidden in his underwear aboard a Northwest Airlines flight as it was preparing to land in Detroit. Suddenly terrorism was back in the headlines. Terrorism and national security had been cornerstones of my campaign: I've had strong opinions about them ever since I joined the National Guard. For months, I had said, "In dealing with terrorists, our tax dollars should pay for weapons to stop them, not lawyers to defend them." But almost the first thing the FBI had done with the underwear bomber was to read him his Miranda rights and then get him a lawyer at taxpayers' expense. And there were hundreds of unanswered questions about why Abdulmutallab—whose name was on federal watch lists and whose own father had gone to the U.S. embassy in Nigeria with concerns that his son had become radicalized by Islamic imams in Yemen and who arrived at the boarding gate apparently without his own passport—was ever allowed to board the plane.

In the face of an attempted terrorist bombing and then a suicide bomb attack on American CIA agents at a

U.S. base inside Afghanistan, Martha Coakley's answer about her foreign policy experience being buttressed by her sister "living overseas in the Middle East" seemed a whole lot weaker and my thirty years of military training and service in the Army National Guard looked a whole lot stronger. And I kept right on talking about terrorism. I was opposed to trying self-admitted 9/11 mastermind Khalid Sheik Mohammed with taxpayers' money in a New York City civilian court, and I wasn't afraid of interrogating enemy combatants under all of our applicable laws in order to discover what other violence they might be plotting against the United States and American citizens. I have always said that the U.S. Constitution and American laws are designed to protect our nation, not to give rights and privileges to people who have not earned those protections, namely our enemies during wartime. And Boston had firsthand experience with terrorism. Two of the hijacked planes on 9/11, the two that hit the World Trade Center, had left from Logan Airport. In the Boston Public Garden, just beyond the fabled duck pond and Swan Boats, there is a small memorial to the people from Massachusetts who died on board those planes.

On Sunday, December 27, my top campaign staff—Eric, Peter, and Beth—gathered to discuss a media buy.

They had prepared one television ad. It began with some old newsreel footage of a 1962 speech by President John F. Kennedy making the case for his tax cut proposal to stimulate the economy, by returning billions of dollars to the nation. Then the old footage faded out, the screen went completely fuzzy, and I appeared, talking about how tax cuts would put more money back into our economy to help create new jobs and new salaries, and those jobs and salaries would create more new jobs and salaries. Lower taxes would equal more jobs. The ad was designed both to highlight my economic beliefs and also to show that decades ago those ideas had been part of the Democratic Party's philosophy as well. One thing that I heard across the state and that my advisers also heard repeatedly was, "This is no longer my parents' Democratic Party. They don't represent my interests." It was one reason why there were so many independent voters in Massachusetts, over 50 percent of the electorate. It was not so much that they had left the Democrats; it was that over the course of fifty years, the Democrats had left them.

We were saving the ad for the last two weeks of the race. But on that Sunday afternoon, the decision was made to release it now, to buy airtime now. Everyone thought we were crazy. The conventional wisdom was that no one would pay attention to the election in the week between Christmas and New Year's Day, that no one would be around watching TV. We settled on a $150,000 media buy, starting December 30. It was a risky ad. I wasn't saying that I was a modern-day President Kennedy; I was saying that his philosophy on spending and taxation was exactly like my philosophy. And what was true then is still true now. Present-day Democrats, especially some in power in Massachusetts, have too often forgotten about that. They don't believe strongly or even much at all in free-market enterprise, in freeing businesses from taxes and regulations to succeed. Their philosophy is more government control, more government intervention, and fewer individual rights. The ad was designed to be daring. We knew people would love it or hate it, but if you believed the *Boston Globe* polls, I was down 30 points. Martha Coakley might want to run a no-direct-debate race and simply coast to election night, but I didn't. The final weeks were going to be anything but a conventional campaign.

The Republican Senatorial Committee got wind of the ad and was horrified to the point of being apoplectic. The members called, telling us not to run it, to take it down. Even Gail was nervous. She said, "People will think it's sacrilegious. You're going to infuriate the Kennedys. You'll get everyone fired up." I said,

"Honey, that's the point. We want people fired up." We did the full media buy. The ad ran for nearly a week. All of a sudden people started talking about the race. Who was this Republican state senator who was linking his views to JFK's? Overnight, the national media began to take note. But what really excited us was the word that the next night at New Year's parties around the state, people were talking with their friends about "the Kennedy ad." Coakley's people wrote it off as a last-ditch effort. And they didn't put any of their own ads on the air. In fact, between Christmas and the New Year, it seemed as if Coakley had taken a vacation.

Now, as with a modeling go-see or a job interview, I was getting a second look. I'd cast over six thousand votes in my time in municipal service and in the Massachusetts state legislature; I'd served in the National Guard for three decades, rising to the rank of lieutenant colonel; I had been a town assessor, a state rep, a state senator; I'd been married for twenty-three years, with two great kids. Both Gail and I had started from nothing, and Gail had a wonderful reputation as a fair, honest, and hardworking reporter. People knew her for her hats and coats and being out reporting in the worst of the winter storms. They remembered Ayla for being a contestant on *American Idol*, how she handled her fame with poise and dignity and became a role model for young girls. Arianna was well-known around Wrentham, at her school, and in our church for her volunteer work and her love of animals. People began to see that I wasn't like the candidates who came out of the Democratic machine, nor could I be written off as a "typical" country-club Republican, and that I wasn't like a lot of other politicians from both parties, a rich guy born with a silver spoon in his mouth. And they knew that I was a hard worker; even people who didn't agree with my politics or my votes couldn't say that I wasn't a hard worker, because I always was. Now more and more of the citizens of Massachusetts were getting comfortable with me, with who I was. They knew that I was tough on fiscal issues and military issues. They also knew that I would listen, keep an open mind, and make an independent decision. I was right where a lot of Massachusetts voters were on the political spectrum.

On January 5, pollster Scott Rasmussen released a poll showing Martha Coakley at 50 percent and me at 41 percent. Her lead was now down to single digits; my gut had the race even closer, tied or with me slightly ahead. That day, I also went back on Laura Ingraham's radio show, this time with Laura herself. She grilled me, and she asked me for my Web site. I worked it in about four times. After the poll and her show, we raised $100,000 online. Now, almost daily, major online money started to come in. That night, we put up our next ad. It featured me in my truck with 199,467 miles on the odometer (now it's got over 213,000 miles and counting), and I talked about how I had been driving my truck around Massachusetts, what I was hearing from voters, and how the truck had brought me closer to the people of our state. The Republican Senatorial Committee balked at this ad too, thinking it had too much about the truck. We put it up anyway, and suddenly I had a new tagline: Scott Brown from Wrentham, the guy who drives a truck. And it wasn't an image; it was the real me, with no deviation. I've had the same truck for years, and it smells like a locker room. Gail always complains about the smell.

I did drive around the state in my truck. And I drove myself. For a while, the campaign tried to get me to use a volunteer aide, but most of the time, the aides didn't drive that well or didn't know where they were going, so I ended up driving them. For a lot of events, I simply drove myself. I drove working two phones, stuffing a piece of pizza into my mouth, sometimes even maneuvering the steering wheel with my knees. I'd hop out of the truck in my barn jacket, the only warm jacket I had, which Arianna had bought for me as a gift, and I'd arrive places alone. People would ask, "Where's your entourage?" And I'd say, "What entourage?" The answer would be, "Well, everyone has an entourage." I'd tell them, "No, I don't—no entourage for me. I'm not Martha Coakley." When she went to events around the state, she had state police drivers and a phalanx of aides and state troopers, plus a gaggle of reporters.

On many days, it appeared as if she had only a couple of public events, and on some days, it seemed that she didn't have any. When she did go somewhere, it was usually a quick in and out, only a few hellos, barely any shaking of hands. She wasn't giving interviews either, and people in the press started to complain. On January 6, Brian McGrory, a *Boston Globe* columnist, wrote a piece titled, "Where's Martha Coakley?" His first sentence was "If you're a registered voter in Massachusetts, your friendly Democratic Senate candidate, Martha Coakley, is sticking her thumb in your eye." McGrory went on to rip Coakley for refusing to debate me live, except once, and for having no campaign schedule to speak of. "For all we know, she's spending much of her time at home with the shades drawn waiting for Jan. 19, Election Day, to come and go." He ended with, "In Washington, senators don't get to dodge their opponents. Right now, dodging looks like the Coakley way."

Our first formal debate, moderated by Jon Keller,

the well-respected Boston political analyst with WBZ, had been taped right before Christmas, and had aired that weekend after the holiday. It was a spirited back-and-forth, and Keller was a great moderator, but because of its airtime early Sunday morning, it did little to change the dynamics of the race. Now, we were scheduled to have another debate in Springfield, Massachusetts, Coakley's home base area, but it was also getting limited airplay and was being sponsored by a small, local public TV station. However, everyone in the room recognized that the stakes were rising. Throughout the debate, she got aggressive, often cutting me off by saying, "Scott, just answer the question. I don't know why you can't answer the question." I'd reply, "Excuse me, I'm trying to answer the question," and she'd say, "You've already answered it." Her whole tone was dismissive, aggressive, and condescending. As I drove home, I thought about it and said to myself: So this is her new strategy. She's going to come out as the big, tough prosecutor, be aggressive, and try to put me on the spot.

Our final debate was going to be held on Monday, January 11, just eight days before the election; it was sponsored by the Edward M. Kennedy Institute for the U.S. Senate and held at the University of Massachusetts in Boston. Throughout the building, there were huge signs with Kennedy's name all over them. We spent our last two hours negotiating to have the Kennedy signs taken off the podiums, but the two of us still stood under a giant sign with Kennedy's name on it. It wasn't subtle enough to be subliminal. It was like a neon sign saying "the Kennedy seat." Longtime newsman, pundit, and presidential counselor David Gergen had been chosen as the moderator. That weekend, the *Boston Globe* released a poll showing me 15 points behind Coakley. That day, January 11, my campaign launched an Internet money bomb. The idea behind a money bomb is to try to reach a donation goal within twenty-four hours. Our goal was to raise $500,000. When we made that number, we upped the goal. In the end, on that single day, we raised $1.3 million from sixteen thousand donors around the country.

It was a bitter cold January night, and outside the hall before our final debate, Coakley's supporters and my supporters were standing around holding signs. I stopped at the bottom of the campus road and walked up the driveway to shake all their hands, including the hands of the guys with the Coakley signs. I told them, "You guys are heroes to be out in this freezing cold. I can't believe you guys. Martha should be very appreciative, and I know I am for you guys just being part of the process. It's been fun, hasn't it?" And everyone on both sides said, "Yeah, it's been fun." And then almost to a person, all the Coakley sign holders, who were mostly union guys, said, "Scott, we're voting for you. We're here because we're getting paid to hold these signs, but we're voting for you." I've been a union guy myself for twenty-five years, and I proudly walked in with their words echoing in my ears.

Inside, I went back to meet David Gergen, whom I'd always admired and enjoyed when I saw him on television. When I said, "Hi, I'm Scott Brown," he looked at me like, "Oh, are you the Democrat? Oh yeah, you're the Republican." He appeared preoccupied, as if he were just there to go through the motions and to get Martha through the debate. The message in the back room was clear: I was going to get crushed—so get ready. Get ready I did. I could tell immediately from the line of questioning what the tone would be: skeptical questions for me, less combative questions for Martha, and partly ignoring Joe Kennedy. The questions to Martha included, "President Obama has made a vow, no new taxes on any couples making less than $250,000. Do you join him in that pledge as senator?" and "There are some who wonder if you have sometimes been a little complacent as a front-runner, although now you're sort of catching fire, and I wonder if now, looking back, you think it was the right decision to insist on three people at a debate?" For me, a typical line of questioning was, "To raise some concerns that have been out there on the campaign trail, starting with you, Mr. Brown. There are those who argue that you've been campaigning as a moderate Republican, more in the Bill Weld mold, but in fact you're quite conservative on some issues." And so I got asked about *Roe v. Wade* and climate change, or asked first whether I would cut entitlements, specifically Medicare, Social Security, and Medicaid. It was a rather disparate style of questioning, underneath the Edward M. Kennedy Institute sign.

One of Martha's and my exchanges came over an amendment that I had offered in 2005 that would have exempted doctors and nurses in emergency rooms from a state law requiring that they offer emergency contraception to rape victims if it violated a "sincerely held religious belief." If that were the case, another qualified medical professional could offer the drug, at no additional cost and with no waiting time, for the victim. It was a measure that Ted Kennedy, a committed Catholic, had supported when he was in the U.S. Senate. In 2009, less than a year before, he had written about his personal belief in a "conscience protection" for Catholics in the health field in a letter to Pope Benedict that was hand-delivered by President Barack Obama. My own amendment was thoroughly debated and received a lot of bipartisan support, but ultimately it was not accepted. However, I still voted for the final bill without the provision, and also voted to override then-Governor Mitt Romney's veto. I started to answer the question and Martha began badgering me over the abortion issue. I turned to her and said, "Excuse me, I'm not in your courtroom. I'm not a defendant. I'd like to have a chance to answer the question." She tried her same line again, and I said, "Martha, I am not a defendant here. Let me answer." It was a devastating counterpunch to her strategy of trying to bait and bully me, like an overeager lawyer in a TV courtroom drama.

We tangled over health care, and at one point, David Gergen jumped in and asked, "Are you willing under those circumstances to say, 'I'm going to be the person, I'm going to sit in Teddy Kennedy's seat, and I'm going to be the person who's going to block it [health-care reform] for another fifteen years?'" For months, all I had heard was that this was the Kennedy seat. Even my wife would sometimes refer to it as "the Kennedy seat." I respected Ted Kennedy. I didn't always agree with him, but I respected him. Still, this was not his seat. And I was getting pretty ticked off. I turned to Gergen and I said, "With all due respect, it's not the Kennedys' seat, and it's not the Democrats' seat; it's the people's seat." And then I kept on with my discussion and my answer regarding health care. As I finished, I looked out into the crowd, and on more than half the faces of the people in the audience was the look of total shock. Toward the front, my campaign team just smiled.

From there, we moved on to terrorism. Twelve days before, there had been a suicide bombing attack on a crucial CIA base in Khost, near the Afghanistan–Pakistan border. Seven CIA officers and contractors, including the base chief, had been killed by the bomber, an al-Qaeda double agent, and six others had been seriously wounded. One of the dead was from Massachusetts. Intelligence officials called it a "devastating blow" to U.S. counterterrorism operations. Gergen asked a question about Afghanistan, and Martha said that al-Qaeda was no longer in Afghanistan: "They're gone. They're not there anymore." I glanced out and my team members looked stunned and started slowly shaking their heads. The people on her team were also shaking their heads and furiously bending over their BlackBerrys. Gergen gave her a chance to fix her gaffe, and instead she put her foot deeper into her mouth. I looked into her eyes and I said to myself right there: it's over; this race is over. That was the moment. The combined impact of "It's not the Kennedys' seat; it's the people's seat," "I'm not in your courtroom," and the statement that al-Qaeda was "gone" from Afghanistan would be too devastating to overcome. Martha and I also sparred over giving terrorists Miranda rights and treating them as ordinary criminals. I said we need to treat them like enemy combatants and interrogate them using our applicable laws to learn whatever we can about what they might be plotting next.

After the debate, I went upstairs to take a quick look at the postdebate news coverage, and the first negative ad against me by Martha Coakley was already on the air. It was a vicious attack ad and a total distortion of my position and my votes, including my views on emergency contraception for rape victims. We had always known that Coakley would go negative. Anticipating that she would unleash a string of attacks, two weeks before, we had taped a response at the kitchen table in my home. It began, "By now, you've probably seen the negative ads launched by Martha Coakley and her supporters. Instead of discussing issues like health care and jobs, they decided the best way to stop me is to tear me down. The old way of doing things won't work anymore. Their attack ads are wrong and go too far." We had our response on the air in a matter of hours.

But it wasn't Coakley's ad that grabbed the spotlight; it was the line "It's the people's seat" that went viral. We had already gotten national media attention.

On January 4, CNBC's Larry Kudlow filmed a segment with me while I was at the Colonial Inn in Concord, Massachusetts, the place where the first shot of the American Revolution—"the shot heard round the world"—was fired back in 1775. Not long after, we started to see homemade signs with the words "the Scott heard round the world." It wasn't quite that dramatic, but four days later, Sean Hannity invited me on for a segment on his major nightly Fox News program, and a week or so later, I was a guest on Greta Van Susteren's show. People across the country were taking notice. And just when we really needed it, money began to pour in.

We raised more money online the day after our money bomb than the day of it, and that was probably nearly all due to the January 11 debate. By week's end, when we got the final tally, we learned that the campaign had taken in $2.2 million in a single day. We didn't want to announce the final figures; we didn't want the Coakley people to know what was coming in and how much the momentum was shifting. The preelection polls were all over the place. Some had the race in a 2–3 point spread; one poll even had me up by 1. In a huge boost during the final weeks, the Republican Senatorial Committee had arrived and set up phone banks with preprogrammed, automatic dialers, which were an enormous time-saver, and they were pulling in volunteers from all over the country. The people who came were excited to make phone calls and it was contagious. Tea Party activists came from other parts of the country to rally behind my message of cutting taxes and returning fiscal responsibility to Washington. We even had Democrats coming over to make calls too. Across the state, in Holyoke, Plymouth, Boston, Needham, Danvers, Worcester, and Wrentham, our phone banks were packed with volunteers. We took over the top floor of a warehouse building in Worcester, and people were standing in line just to be able to take turns to make calls.

I received the endorsements of the State Police Association of Massachusetts and a number of local police unions, two in Worcester—the New England Police Benevolent Association Local 911 and the International Brotherhood of Police Officers Local 504—as well as the Cambridge Patrol Police Officers Association, which was the union that Martha Coakley's husband had belonged to when he was a Cambridge cop.

In those final weeks, I also got some unexpected help from the *Boston Globe* and some of the Boston TV stations. Every *Boston Globe* poll that came out had Coakley vastly ahead, which in our view, inside my campaign, only served to make her supporters complacent and to hide the level of real trouble she was in. And others in the local media also couldn't imagine that I could win. Over at Channel Five, Gail's station, the regular on-air political consultant Mary Ann Marsh had already concluded that the race was over. When the Kennedy family endorsed Martha two weeks before the election, Mary Ann kept up her negative barrage and called it the final blow to my campaign, noting that Martha Coakley had received an extra $100,000 in online donations as a result of that endorsement. What she and others didn't bother to pay attention to or to report was that we had $100,000 coming every hour or two on the days surrounding our money bomb. And we had money coming in after that, checks that were for $25 and $50; our average donation was just $88. I had people showing up at headquarters who would say, "I've driven in from Winthrop," or "I've driven in from Sturbridge"—or Pittsfield—"to bring you this check." People dropped off $1,000 checks and all they wanted in return was a smile and a handshake. And whether people gave $5 or $1,000, we treated them the same.

Coakley had her attack ads and her endorsements and was rallying Democratic leaders. As the *Boston Globe* put it in a profile that ran on January 13, "she had little time for the hand-shaking and baby kissing of a standard political campaign." On January 12, the day after our final rough-and-tumble debate, she didn't make a single public appearance, and instead flew to Washington, D.C., for a fund-raiser, to try to get a big infusion of special interest money for her campaign. Of the twenty-two-person host committee for the fund-raiser, about fifteen of the hosts were federally registered lobbyists with health-care clients.

But that was only part of her campaign's strategy. For weeks now too, the Democrats had been following me around with "trackers," people who filmed my stops and my events, trying to catch me making some gaffe. I would see them riding behind my truck in their cars. I easily recognized the main tracker, and I used to say hi to him during almost every stop. I'd say to the crowd, "Excuse me, I would like to introduce you to the tracker for the other candidate. The other side actually has someone following me around everywhere I go. I see him first thing in the morning. I've seen him follow me in a car. I just want to say, 'Thank you very much, Mr. Tracker, for being so respectful and courteous. You've been a fantastic tracker. I'm just hopeful that at the end of the race, maybe I could get copies of what you have, so I can watch the videos when I'm old and I retire.'" My tracker worked for the Democratic National Committee, and he was a nice guy who laughed when I made my jokes. They were looking to catch me, just as they had tried to catch me in the fall when the race started. During the election, they sent teams to scour my legislative and municipal voting records. They combed through town hall to see if I went to town meetings, if I had paid my real estate taxes, if I had registered my cars, if I had even gotten the proper licenses for our two dogs, Snuggles and Koda.

Now they were trying everything. They sent out actors to my campaign events who dressed as Wall Street millionaires and walked around carrying glasses of wine and champagne, shouting taunts and attempting to harass me, trying to tie me to a bailout that then-Senator Barack Obama had voted for and to imply that I was backed by Wall Street tycoons. On the podium, I would simply say, "Hey, guys, Halloween was months ago," and proceed to ignore them. We put together what I called the truth squad, the Brown Brigade, of people who monitored Web sites and wrote letters to the editor to counter the often vicious and wrong things that were said about me and my campaign. We also had a way for people who supported me to link up via social networking.

Meanwhile, I had been driving my truck into South Boston.

South Boston is best known for being a working-class, largely Irish-American neighborhood of solid, unadorned row houses and small businesses, mom-and-pop places, old-style barbershops, and luncheonettes that still serve regular coffee in cream-colored stoneware mugs. I would go there early in the morning, when the commuter traffic was coming in, and I'd stand in the four-corners area off Broadway and hold my large "Brown for U.S. Senate" sign, a cup of hot chocolate in one hand, and I'd wave at the people who drove past. But not everyone was passing. People were stopping their cars to shake my hand; they were giving me thumbs-up; they were asking for bumper stickers; they wanted signs; they brought me coffee and hot chocolate. They offered their help. I would come back to our small offices and tell my team that I thought we were doing really well in South Boston. My campaign advisers weren't big fans of going out and holding up signs, but after the third time I went out there, they were curious and they came along with a video camera. It was just how I had described it—people coming up to me, engaged, excited; my being mobbed to shake hands. I spent some other mornings in front of the South Station train station in Boston, greeting commuters coming in from all over the state, and the response was phenomenal.

By the last week of the campaign, we didn't have enough signs and bumper stickers to give out, so people in and around Boston and the rest of the state started making their own. They Scotch-taped handmade signs to their car windows and on their trucks, even making homemade floats that they pulled along the road, all to show their support. Some people even took our signs off the front lawns of supporters and moved them to more high-traffic areas—or to their own houses.

I had picked up endorsements from Steve DeOssie and Fred Smerlas, formerly of the New England Patriots, and was thrilled to earn the backing of both the great football quarterback Doug Flutie and Curt Schilling, the famed Boston Red Sox pitcher who had played with a bloody sock in Game Six of the 2004 American League Championship series against the Yankees. His white sock soaked with blood, Shilling pulled in the win. It sent the championship to Game Seven, and ultimately earned the Red Sox a place in the World Series, for the first time in nearly twenty years. Schilling pitched with a bloody sock to help win that series too.

Martha Coakley's reply was to dismiss Curt Schilling by calling him "another Yankee fan." Live, on the radio, on Dan Rea's show. When Dan, sounding completely incredulous, asked, "Curt Schilling, the great Red Sox pitcher of the bloody sock?" her reply was, "Well, he's not there anymore," a statement that was completely unbelievable to almost anyone from Massachusetts. She also mocked me for standing outside in the cold to shake hands with people going into Fenway Park for the Winter Classic. "In the cold? Shaking hands?" she told the *Boston Globe.* "This is a special election. This is about getting people out on a cold Tuesday morning." What she did do was tell a meeting of ten local officials, "There is no way in hell Massachusetts is going to send a Republican to Washington."

While Martha brought Bill Clinton in to campaign for her, I went to the historic North End of Boston, home to many hardworking Italian-Americans, and hosted a rally with Rudy Giuliani. With help from the Riccio family, Joe Ligotti, and Peter Marano, we had an incredible turnout. It was standing room only; there were so many people that the entire street was closed. The feeling was overwhelmingly uplifting, and that was the type of race I wanted to run. When Republican entities or outside special interests put up negative ads, I told them to take them down. I wasn't going to go that route. I had committed to running a positive campaign, to talking about the issues, to not making negative, personal attacks. Martha Coakley and the Democrats went completely negative. Some Democrats sent out a mailer, based on the UPS slogan, "What Can Brown Do for You?" It attacked me and my positions. UPS was furious.

The Coakley team used negative calls, negative polling. They put out one ad on abortion, saying that I'd turn away rape victims from emergency rooms, and another mailer that was so blatantly false we held a press conference with our legal counsel and threatened to sue under the Massachusetts statute that makes it a criminal offense to knowingly make a false claim in a political race. After the election, the chairman of the state Democratic Party personally called me to apologize for the mailer.

Ayla and Arianna were distraught listening to the attacks. They knew what type of dad I was. Gail couldn't say anything, but they could. They asked how they could help. They came out to hold a press conference and speak to reporters denouncing what was being said about me, adding that such lies are why young people are discouraged from going into politics. They asked me if they could tape a radio commercial too. It was possible for anyone listening to hear the hurt radiating from their voices.

Meanwhile, I got on a bus and traveled around the state, holding rallies where thousands of people came out to stand in the cold. When I talked about the ads, I would just say, "Shame on Martha," and the crowds would start chanting back, "Shame on Martha, shame on Martha!"

The final weekend before the vote, on Sunday,

President Obama flew in to try to salvage the Coakley campaign. In his speech to energize her supporters, he mocked my truck, saying, "Anyone can buy a truck," and "I'd think long and hard about getting in that truck." The next night, I returned home to a rally in Wrentham. Ayla and Arianna were with me, and I got up and said, "Mr. President, you can criticize my record, you can criticize my policies, you can criticize my votes, but don't ever start criticizing my truck. With your help, I'm going to gas up that truck and drive it right down to Washington." The crowd went wild, chanting, "Gas the truck, gas the truck!"

On Election Day, just under 2.25 million people went to the polls to cast their votes—earlier commentators had expected a turnout of my old best guess, around 600,000. That night, some of the local television stations sent their top reporters to Coakley's headquarters and their B-teams to mine. We knew what was coming, even if they didn't. A little over an hour after the polls closed, Martha Coakley called me to concede. The national media called the race by 9:30 that night. I had won by 5 points and 110,000 votes; according to the finally tally, 25 percent of the Democratic vote had gone for me. Eric, Peter, and Beth like to point out that they never had me practice a concession speech.

Of course, the start of my victory speech was unrehearsed. It was an exciting moment, everything was going great, and as I was thanking Gail and Ayla and Arianna, I made another one of my bad jokes, saying that Ayla and Arianna were "available." It was a joke that only a father understands, and only a devoted dad could actually make and occasionally get away with it. My daughters are never available. They know I love them and would do anything for them. They also realized when I cracked one of Dad's bad jokes that I was back; I was their dad again, and no longer in campaign mode. It was a joke in the same way that my high school buddies call me up and say, "So, Senator Brown, can I still call you a loser?" and proceed to do just that.

It was crowded that night on the podium. My mother and father were there, and so were Leeann, and Robyn and Bruce. I had some of Gail's family, my nieces and nephews, cousins, so many people who had thankfully circled back into my life. I never thought I'd be standing there, surrounded by all these parts of my past, but I was. For the first time in my memory, everyone was smiling.

And for the first time, on election night, in public, I had Gail. Channel Five initially hadn't wanted her to appear at all. Then the channel came up with all kinds of restrictions: she couldn't answer any questions; she couldn't be seen in any photographs that might be disseminated or used for publicity purposes. We finally agreed that she could only appear with me in public after one side or the other had conceded the race. On the night of my primary win, she had come with me to the hotel, but she had to remain alone up in the suite while I went down to give my victory speech. When reporters came to the suite to ask me questions, Gail was forced to hide in the bathroom and pretend she wasn't there. She had to spend that evening by herself, in tears, while I was downstairs celebrating with our kids, my supporters, and our friends. Now I finally got to put my arm around my wife on election night and celebrate a win, in public.

That night, I remembered Ted Kennedy for his service and spoke of how "I honor his memory" and pledged myself to be a worthy successor. I spoke of the people I had met in every corner of the state. I looked them in the eye, shook their hand, and asked for their vote. I didn't worry about their party affiliation, and they didn't worry about mine. It was simply shared conviction that brought us together. I promised to work to put government back on the side of the people who create jobs and on the side of the millions of people who need jobs. And I spoke about working on behalf of our veterans and working to keep our country safe. I ended with the lines, "I've got a lot to learn in the Senate, but I know who I am and I know who I serve. I'm Scott Brown. I'm from Wrentham. I drive a truck, and I am nobody's senator but yours."

Chapter Seventeen

MR. BROWN GOES TO WASHINGTON

In August 2005, Ayla auditioned for *American Idol.* Gail and I weren't eager for her to do it. Gail was worried that she'd be devastated if she didn't make it through. I was worried about what would happen if she did. When she was only fifteen, Ayla had been recruited to play basketball for Boston College, with the offer of a full scholarship and a spot on a great team. She was one of the youngest recruits to commit to BC in the school's history. Now, at age seventeen, she was entering her senior year of high school. *Idol* would require her to be in California, to miss games, to miss starting work with her Boston College coach, Cathy Inglese. But she had always loved singing. She had persuaded us to let her take singing lessons when she was sixteen years old. She had already been in some local competitions. Some she had done well in, others she hadn't, but she had a great voice. The auditions were being held at Gillette Stadium, where the New England Patriots play, in neighboring Foxboro, five minutes away. We relented and said yes.

The days of the auditions, it poured rain. Not a little rain, but massive, drenching downpours. Ayla and Gail waited in line for eighteen hours. I kept driving over with blankets, food, and dry clothes. We'd hold up the blankets and Ayla would change right there, waiting in line. There were six stops at the audition table. Each hopeful had to sing for about fifteen seconds in front of a producer. Ayla had chosen "Ain't No Mountain High Enough," and she kept getting ushered through to the next producer and the next round. Then the auditioners asked to send a crew over to our house to film. I had a feeling by then that it was pretty likely that she was going to go all the way, to sing in front of the judges' panel, Randy Jackson, Paula Abdul, and Simon Cowell.

In her final audition, Randy and Paula, but not Simon, gave her a golden ticket and sent her through to Hollywood. That final night of tryouts, she got to sing the national anthem before the New England Patriots' game, which *Idol* also filmed. We knew that she was moving through, but the mammoth contract she had to sign meant that we couldn't tell anyone that she had made it to the next level of *Idol*. Everything had to be kept under wraps until the show aired.

Hollywood week was pretaped. Gail took Ayla out to California and didn't tell a soul, except her basketball coach. Because she was only seventeen, either Gail or I or another family member had to be with her. Gail had the first shift. The days were grueling. Added to that is that all of the younger contestants, those under eighteen, have to attend four hours of school a day to comply with California law, so while the older singers have a chance to sleep, the younger competitors are in the classroom. Ayla made it through Hollywood week. When Gail had to be back at work, I took over; then Gail's sister, Ayla's aunt, flew out for a few days. In between, we were flying Ayla back and forth to compete in her basketball games. If she suddenly stopped showing up for basketball, all the secrecy and confidentiality that the show required would be gone. Ayla would fly in on the red-eye, throw on her uniform and go play, and then turn around and fly back out. And she managed to be the high scorer in nearly all the games.

Her last game was in Rhode Island and she scored 40 points; she was 10 for 10 from the foul line. Afterward in the locker room, she broke down in tears, telling her teammates that she had to leave them to go compete on *Idol.*

For all the things that you see when the show is boiled down to an hour or so of tape, behind the scenes everyone at *Idol* was very professional and very nurturing. Paula Abdul in particular was wonderful to Ayla, and the show has not been the same without Paula.

Ayla made it to the top twenty-four, then the top twenty. Then the next week, in the round of sixteen, which would pare the contestants down to the top twelve, she sang British singer Natasha Bedingfield's "Unwritten." I arrived just before the results show started and found her curled up on the couch with a high fever. The rooms were cold, particularly the hotel rooms, and she had throat problems, combined with the bad food and overall exhaustion. She was simply worn-out. She didn't make the cut to the top twelve, although Chris Daughtry, Taylor Hicks, Kellie Pickler, Ace Young, and Mandesa all did. We still keep in touch with Ace, Kellie, and Mandesa. I was there for results night, sitting in the audience, watching her cry as she sang her song one last time. It was the most helpless I had ever felt as a parent, watching my baby cry, and having to grip the arms of the seat not to just rush up onto the stage and hug her.

Backstage, all the emotion caught up to her, and Ayla, who is like me and almost never cries, was bawling her eyes out. Paula Abdul came up to give a hug and say how sorry she was. I thanked Paula, but I said that I thought the show had it wrong about Ayla. She wasn't a girl from a privileged background, as the TV clips had made it out to be. Her mom and I had both come from nothing, and she had worked herself for everything she had. At home, her car was a used 1983 Crown Victoria. But one thing did come out of those two nights: Natasha Bedingfield's "Unwritten" vaulted to number one on iTunes. Ayla met Natasha the following summer when they both performed at the KISS 108 radio station concert outside Boston, and Natasha gave Ayla a big hug and thanked her for helping to make the song so popular.

Ayla turned her thirteenth-place finish into a great career all through college. She went on to cut a couple of CDs and to sing in all kinds of venues, including for the Celtics before a number of their final playoff games, so much so that they've considered her their lucky charm. And of course, when she came home, she started training for the Boston College basketball team. The toughest challenge for Ayla was trying to balance the NCAA restrictions with also having a career. She did it all four years without incident. We all regretted that Ayla did not go a bit farther on *Idol*. Arianna had a ticket to come out to Los Angeles the day after Ayla was eliminated, and Gail and I are still sad that Arianna never got to participate in her sister's *Idol* experience in California.

So, before January 2010, I had always joked that I was tied at a distant third with Arianna among the most well-known people in the Brown household, after Gail and Ayla. I had learned a great deal from them, watching my wife and my daughter handle the media spotlight with grace and aplomb, and watching Arianna handle the high pressure of horse meets, where she earned bright ribbons for her skilled and elegant competition. But with the vote on January 19, the level of outside interest in the Brown household was now about to radically change.

I'm grateful to my family for helping me to be prepared for the full deluge of press scrutiny. Had I come from an entirely private life, it might have been totally overwhelming. Now, it was just partly so. On one of my first plane flights down to Washington, a local reporter showed up at the arrival gate and was waiting for me as I got off the plane. I recognized him and asked him if he was going on a trip, and he told me that he had bought a travel ticket just so that he could get past security to talk to me when I landed. I told him I hoped that it was going to be worth the money he'd spent.

Everywhere I went that day, I had cameras in my face. Reporters surrounded me with microphones and lights like a swarm of buzzing bees. After the election, I had to fight just to get sworn in. The Massachusetts secretary of state wouldn't immediately certify me, although the office had done it in three days for Niki Tsongas, the widow of former U.S. Senator Paul Tsongas, when she won a special election for Congress in 2007. But this time, state officials apparently felt that it was imperative to wait ten days for overseas absentee ballots to be opened and counted. I had to work to get them to certify my win and to make sure that Vice President Joe Biden wasn't going to be away or the Senate out of session so that I could be sworn in. Biden had called me on my cell phone on election night, saying, "Scotter, this is Joe, Joe Biden, the vice president. I guess I should call you senator. You ran a hell of a race. This is Joe Biden, the vice president. Hell of a race. Congratulations." I also got calls from the president and other top leaders on both sides of the aisle.

Vice President Biden swore me in on February 4, the day before a two-pronged epic snowstorm began to hit Washington, shuttering the federal government. A month or so later, we had lunch at the White House. He gave me a tour of the West Wing, from the Situation Room to the Rose Garden and the putting green. The day before, President Obama had signed the massive health-care overhaul bill, and Biden had been caught on a live mike telling the president, "This is a big fucking deal." As I finished my tour and sat with Biden in his bright blue West Wing office, with a fireplace at one end and portraits of John Adams and Thomas Jefferson, the nation's first two vice presidents, on an adjoining wall, I said to Biden, "Mr. Vice President—" He promptly interrupted and said, "Hey, Scott, call me Joe." "Joe," I said, "this is a big fucking deal." He looked at me, grinned, and said, "You're a wise guy, senator. I like that. We're going to get along just fine." The lunch was scheduled for thirty minutes, but we spent over an hour and fifteen minutes eating, until his staff forced us to finish up. I gave him an Ayla Brown CD and a Red Sox hat with the number 41 on the side, for the forty-first senator, as a parting gift.

Back in Boston, Gail took vacation time from work. I packed up some things and readied my truck, and we drove down to D.C., in time for the second wallop of snow on February 10. I had to start hiring a staff and doing basic things like getting computers, getting the Senate office painted, and locating furniture in the aftermath of a crippling blizzard. What most new senators have two or three months to accomplish, I had to do in all of one week. However, I had the upper hand. I was a Boston driver behind the wheel of a big green pickup truck on the streets of Washington during a snowstorm. Asidc from a few plows, I was one of the only people able to drive on the roads.

Almost from day one, Senate Minority Leader Mitch McConnell called me and kept in close touch. He was welcoming and generous with his counsel, while always recognizing that I had pledged to be my own man. But Senate Majority Leader Harry Reid also reached out. His first call came on a Sunday morning. When I asked him where he was, he told me he had just stepped out of church to make the call. He asked me to look with open eyes at a bill he was proposing. I said I would, and then suggested that he go back inside to church. It was a jobs bill. I knew it would be a deeply controversial vote for me to take office and promptly side with the Democrats on a piece of legislation. But I felt that I owed it to my constituents to read the bill. What it proposed was targeted tax cuts for employers, designed to create more jobs. Its centerpiece was a payroll tax cut to take some of the burden out of hiring, and its price tag was $15 billion. I knew that there were more potentially divisive and expensive amendments waiting in the wings, totaling upwards of a hundred billion of extra dollars that we could not afford. I decided to cast the deciding vote for cloture, which in the U.S. Senate means a vote to end debate and move the legislation to the next level, usually toward a final vote on the measure. I thought that given the dire economic circumstances of so many Americans, Washington needed to focus on jobs, and on ways to improve the job picture without adding to the already crushing national debt. I didn't want to play politics and just score some short-lived political points. Sadly, after that one bill, the issue of jobs largely vanished from the Senate's agenda for most of the rest of 2010.

Whenever I went to vote, gaggles of reporters stood in the hallway outside the Senate, asking me what I was going to do. The first few times, I felt as if I had been transported back to infantry basic and was standing in the rocket simulator, with explosions detonating above me. Instead, this time, it was the blinding lights and camera flashes from the mass of press.

One night in Washington, to relax, Gail and I walked into the Woolly Mammoth comedy club, only to find that four of the skits in the show were about me; we had already seen the very funny one on *Sat-*

urday Night Live, a takeoff on the *Cosmo* guy comes to D.C. That night at the Woolly Mammoth, people in the audience asked if I was offended. I said no, I have a sense of humor, something a lot of people in Washington need to have more of. Afterward, a bunch of the cast asked to meet me and take photos outside in my now famous truck. I'd now go to Costco or Home Depot and people recognized me. My office was sent fan mail and requests for pictures. But we also got death threats and mail from people who threatened to do us bodily harm. For a while, I had to hire armed guards to patrol my home in Wrentham. If I walked our dogs, someone with a camera would follow me to make sure that I picked up the poop.

For the first month, I rented former Senator Fred Thompson's place, but then for a while, I just slept on the couch in my office. We were so overwhelmed with hitting the ground running that I was always at work. My staff was stunned that I would crash on my couch, but to me, it was like a throwback to the days when I had moved from apartment to apartment, my stuff wrapped in a few blankets, my clothes in a small suitcase. Although I wanted to do business in a new way, I was also very conscious of keeping what worked. We kept a couple of key Kennedy staffers: a lovely lady, Emily, who had adeptly handled constituent issues;

and Larry, who had for years run the mailroom. Good people are just good people. And there are some things about U.S. Senate work that are completely nonpartisan. One of the first things my staff and I did was to help a Massachusetts family whose daughter, Britney Gengel, had been killed in the disastrous Haiti earthquake. They wanted to locate and identify her body, but massive amounts of red tape had intervened. We worked with Senator Kerry's office and Congressman James P. McGovern to help them bring her remains home.

But for years now, Congress had become increasingly partisan. The 60–40 vote split that had prevailed for over a year in the U.S. Senate had left most people on either side completely focused on the numbers. When I sat down to lunch with Joe Biden after my swearing-in, he told me that the Obama Administration couldn't add one amendment to health care, "We couldn't make one change because of you, Scott. We couldn't accept one amendment because you could stop the entire health-care bill, and we needed a victory." I told him I thought it was a deeply flawed bill, one that would hurt seniors, hurt veterans, hurt millions of people around the nation. "It's a bill that's crushing Massachusetts, ruining medical device companies, killing Medicare, and costing jobs." It will likely lead to longer lines and less coverage. And it will likely worsen rather than improve care. I've long believed that all Americans deserve health care, but we shouldn't have to raise taxes and expand the federal government to provide it. I asked him why they didn't just start over. Because, he conceded, what they wanted was the political win. Biden said, "We'll fix it. Don't worry. We know. We know. But it's because of you."

The problem with the health-care bill is that it represents total government domination of an industry, and it was passed by pushing it through with a parliamentary maneuver called reconciliation, which showed a profound lack of respect for the wishes and the will of the voters.

And a lot of the bill isn't as advertised. We've had hundreds of residents in Massachusetts calling my office about provisions that aren't going to be available for three or four years. I tried to get health insurance for Ayla when she graduated from college, because the act is supposed to allow kids up to age twenty-six to use a parent's policy, only to find that the provision didn't take effect with the bill; it takes effect afterward. It is these types of games that have gotten the rest of the country very frustrated with Washington.

It's the same thing with too many other pieces of legislation. It was openly said around the Senate that when now-retired Connecticut Senator Chris Dodd told President Obama that he could craft a bipartisan financial regulation bill that could get 70 or 75 votes in the Senate, President Obama told him all he wanted was a 60-vote bill. He didn't care if they got many Republicans. In fact, he wasn't particularly interested in getting them. In Massachusetts, where in the state senate there were only five Republicans out of forty senators, that might make sense. But in Washington, that isn't representing anyone. Instead, the idea was to make this bill political, to be able to tag Republicans as supporting Wall Street and Democrats as supporting Main Street. It was done partly for political points and political advantage, for an issue in the November elections.

And in the intervening months, the political gamesmanship only accelerated. Throughout the summer of 2010, the Democrats brought up bill after bill, but they did so in a highly partisan way. Through parliamentary maneuvering, they refused to allow Republican senators to offer any amendments or to have even basic input. The whole point of the practice is to try to claim that Republicans are obstructionist, when the real fact is that they have frozen one side, the Republican side, out of the process. This is not solving real problems; this is political posturing and pandering, and it is only compounding the very serious problems that we face as a nation. And we have to face them together: they are our problems, regardless of party. We can't solve them unless we all work together. But too many Democrats in leadership positions have not been interested in doing that.

And it isn't always better from the other side. When I cast a deciding vote in favor of the jobs bill being pushed by Harry Reid because it contained an employer tax cut, conservatives pilloried me. But I don't see my job as being obstructionist all the time; I see it as helping to get Washington moving again. And I think people on all sides have to appreciate that one of the functions of a U.S. senator is to work on behalf of the men and women whom he represents. I'm going to be true to my principles: I'm a staunch fiscal conservative, a committed tax cutter, tough on national security, but if you're looking for someone who is going to be a full-on ideologue always marching in lockstep with his party, I'm probably not your guy. What I've always believed is that there are good people on both sides of all the issues, and that we should listen carefully, have a respectful debate, and find common ground where we can.

I come from a smaller state, but often I think that American political life has become too small. Increasingly, politicians are tempted to focus on petty personal attacks, on scoring political points, and on scolding the very citizens who elected them into office. We as a nation can and should do better. There are heroes in public life, and we would all do well to consider them. I think of General Dwight Eisenhower, who after helping lead America and the Allies to victory in World War II, stepped up to lead the nation in political life, bringing with him a strong commitment to his country and a steady hand. I also think often of the man who shaped my youthful decisions and still deeply inspires me, Ronald Reagan. He possessed confidence, uncommon vision, and most of all optimism. When critics mocked him for his deep faith in America and in the American people, he simply smiled and believed all the more. He lifted up our nation from the depths of economic despair and never faltered in his belief that freedom was better than tyranny. He did not ignore dissidents languishing in Soviet prisons; he gave them his support and he gave them hope. He spoke bravely for millions who yearned for a better life when he said, "Mr. Gorbachev, tear down this wall." He believed in our country at a time when it was all too fashionable to doubt America. He was a uniter, bringing people together around a shared philosophy and shared goals. Nothing says that more clearly than the phrases "Reagan Republicans" and "Reagan Democrats."

There are still examples of this spirit of giving back today. I also think of two men in particular with whom I have had the pleasure of working. One is Mitt Romney, the highly successful businessman who decided to enter public life and to give back. He could have enjoyed his private success, but instead he wanted to share his understanding of fiscal and economic issues. He was willing to do more than he had been asked. And today, in the U.S. Senate, I am honored to serve alongside John McCain, a true war hero. John McCain could have used his family connections to leave his imprisonment in Vietnam, but instead he remained with his fellow Americans. He has devoted his life to public service and to his country. He has always looked to the greater good.

I also came to Washington with a few examples of that myself, starting with Jo Ann Sprague, the Massachusetts state representative whom I replaced. Jo Ann entered politics after serving in the military in World War II and then raising her family. While I was a Wrentham selectman, she acted as my mentor and was always someone I could call with a question. She embodied the philosophy of term limits in how she conducted her political life, and her integrity was a key lesson at the beginning of my political education. Along with Jo Ann, I benefited enormously from my time in office with the four other Republican Massachusetts state senators, Richard Tisei, Robert Hedlund, Bruce Tarr, and Michael Knapik. The five of us called ourselves the band of brothers. Our small numbers made every day a constant battle, but we were determined to make a difference, not only to represent the people who had elected us to office, but to try to improve our state as a whole. We believed in the power and the free exchange of ideas and the need for honest discussion and open debate, something I've tried to carry with me to Washington.

I think today we can do better at all levels of government. I am proud that the January 2010 Massachusetts special election inspired candidates around the country to run for seats the following November. Would-be candidates and volunteers who might otherwise have sat on the sidelines got involved in the process. That is what we need to have a vibrant democracy. If my run for the U.S. Senate helped to motivate them, I am deeply proud of that. I'm glad too that so many races across the country were competitive, in some cases for the first time in years. I personally spent many weekends traveling the nation to help dozens of candidates, making rally and event appearances with them, and working every bit as hard on their behalf as I did during my own race. Many of the people running for office in 2010 were also first-time candidates for public office. They prompted debates and discussions that have benefited all of us. But after this most recent election, the time has come for the country to look forward.

When I came to Washington, I made a "no earmark" pledge. I didn't need a piece of legislation or a Senate rule to enforce it. I made it to myself, and I made it because it is common sense. I hope as the next years unfold, both sides of the political aisle will learn from past mistakes. We do not need a dysfunctional government; we have had enough of gazing in the rearview mirror and assigning blame. What we need is a government whose top priority is to get the economy moving; spur job creation; eliminate uncertainty for business owners, families, employees, and entrepreneurs; and make sure that tens of millions of hardworking homeowners are not trapped in the housing crisis.

We need to work together to solve problems. We have been sent not to relive the past, but to work together toward a better future.

On one of my early trips back home to Massachusetts, I was walking through the airport when a woman from US Airways came rushing out of her work space. She came up to me and said that she knew the whole New England delegation. "I hope you're not going to be one of those nutty Republicans," she told me. I just smiled.

A little over one hundred days later, in the middle of June, the same woman vaulted out of her spot and said, "Come here." I did, and she gave me a big hug. She told me that she thought I was doing a great job and asked me for my autograph.

I like to think that the job I've done is the job that I was hired by the people of my state to do—to go to Washington to represent them, to read each bill and think: How does this impact the families who sit around their kitchen tables with a stack of bills deciding which ones to pay, and how does this impact the people who work at two and three jobs to give themselves and their families a shot at a better life? In Massachusetts, with our appallingly high taxes and housing and heating costs, you can make $150,000 a year and feel like you're barely getting by, especially if you're trying to save for college or to help care for an elderly parent or relative. While it may seem politically advantageous to try to peg Republicans as the party of "no," I'm happy to be guilty of saying no to higher taxes, out-of-control spending, and overreaching government.

I've always thought that every politician ought to have held a real job, and held one recently. It would be good if most of them could be self-employed. More of our officeholders need to know what it's like to wake up in the middle of the night worrying about whether you can make your payroll, whether you can properly compensate your employees, pay the rent, pay the taxes and fees, break even, or make a profit. I've done all of those things for my private law practice. If politicians had to worry about where their money was coming from, maybe they wouldn't be so cavalier with all of our tax money, and they'd give a second thought to how hard those taxpayers have worked to earn what they dutifully send in each week or each quarter of the year.

One of my most gratifying accomplishments as a state legislator was helping the town of Needham build a new high school. The town was caught in a rule change in how schools were supposed to be built and funded, a change that would have cost the town and its residents an extra $15 million. I worked with Senate President Travaglini to exempt Needham from the rule and save the money. The unions weren't happy, but I saw no reason to burden the town residents, who were already paying an enormous amount to try to give their kids a better place for an education. In Washington, the numbers are bigger, but the issues and games are the same. We aren't looking hard enough for ways to save the money of all our citizens. We can do better.

I still drive myself around in my truck. It now has over 213,000 miles. In the winter, I wear the same brown barn jacket. I miss biking and running all the time, but I've become a gym rat in the Senate gym. I've tried to get to know my colleagues. Most are incredibly dedicated and hardworking, more than most people might think. I've built some great relationships with many senators on both sides of the aisle. While remaining true to our convictions, we are all gratified when we can find common ground. Of course, there are a few senators who are punching the clock until the next election, but not a lot of them.

I finally gave up sleeping in my office and eating cafeteria food for every meal; I didn't go to Washington to try to shorten my life span. I now have a four-hundred-square-foot place two blocks from my office, and it feels like a college dorm room all over again.

All this has been hardest on my family. I don't get to see my daughters as much as I would like. I'll never get over missing Ayla play in her last senior game at Boston College or missing Arianna's sorority's family event. I'll take their calls on my cell phone before I'll take one from any politician.

My wife, Gail, has borne perhaps the biggest burden. During the campaign, there were countless rumors that she was privately coaching me, giving me TV and teleprompter training, grilling me on possible questions, working on my statements and my commercials. Nothing could be farther from the truth. The only thing she weighed in on was the Kennedy ad, and she didn't like it. We ran it anyway. Gail and I hardly ever discuss politics. When I showed Gail and the girls the truck ad, she didn't like that one either. We ran it, and it was a huge hit. Afterward, she shook her head and said, "I guess I just don't know politics." It became a running joke in our house and with my small campaign team: if you ask Gail's opinion, and she says do it, then we do the complete opposite.

But the challenges for her magnified after I won.

Now we can't sit down to eat in a restaurant around Massachusetts without being interrupted. I enjoy meeting all the people, but she'd like a little privacy. Even at the end of one of my triathlon races, people stop us for photos and autographs, and Gail becomes their photographer. She does this with grace and good humor and makes everyone feel special, as only she can. Still, for the moment, the peace and anonymity we once enjoyed are very much gone. When we were invited to the White House correspondents' dinner, I thought Gail looked beautiful, but her dress was ridiculed on the Web. It was deeply hurtful to me and to her. It was a level of scrutiny of a spouse that I can't understand.

When Gail went back to work in the news department of her TV station in Boston, her e-mail inbox was deluged with requests for me—people looking to contact me for help with some problem—or with criticisms of me, or with messages from people just sounding off on some issue. They went to her instead. She spent hours replying that they needed to contact my office and giving them the contact information. Once, when she was out on a story involving a suspected homicide, a local police representative came up and began yelling at her, demanding that she tell me that the police need to get more federal grant money, even though I had been in Washington for only a couple of weeks. It's been hard for her, after all those years of working so diligently on being objective, of doing her job while I did mine, to suddenly be caught up in my new profession. But the hardest part was the separation. When I was a Massachusetts legislator, I didn't have to spend my weekdays in another part of the country; each night, I came home. Gail and I both suffered under the separation. Our girls are grown and out of the house; we are empty nesters; and with my new responsibilities, each of us had to make the transition largely alone. Wanting to be together, in the summer of 2010, Gail made the very difficult decision to leave Boston for now and go to work part-time for a local Washington, D.C., news station, covering stories that have nothing to do with politics. She's the one who has made the greatest sacrifice, who has upended her life for me.

I wouldn't be honest if I didn't say that the pressure of national politics is intense and thoroughly unrelenting. But then I think to myself: I could be a spectator in the nosebleed seats, or I could be down on the court in the middle of the game. I've always chosen the court. If there's a battle, I want to be in it, diving for the ball, for my team, ready to take the last shot.

There is a core set of beliefs that have shaped my thinking since the beginning of my service in public life. They are uncomplicated and straightforward. I believe that government is getting too big. We have too many laws, and we don't need more of them. I have always felt that we need to get rid of some of our laws and streamline the ones we keep, and we need to better enforce the laws that we have. Some legislators like to trumpet all the bills that they have introduced and passed, but I like to start by improving what we have, by fixing problems and offering amendments to make things clearer, simpler, and better for the people who have to live with the law's effects. Expand our economy and defend our jobs; don't expand our government and defend wasteful spending.

I believe strongly that taxes are too high. But they are now getting much worse. For months in 2010, Congress and the administration dithered over taking still more money out of the pockets of hardworking Americans. This is an issue that deserves serious attention, but rather than address it and give American families and businesses confidence that they will not face dramatically higher taxes in the coming years, which would have helped the economy emerge far sooner from a terrible recession, the House and Senate spent the overwhelming majority of their floor time on a health-care bill full of new mandates and taxes—like the new tax on medical devices—legislation that a majority of the American people did not want. Even after the health-care bill was passed, very little effort was spent on thinking of ways to revive the economy. Instead, hardworking Americans are likely to be stuck with bigger government bills.

Only after the Republicans won control of the House of Representatives by a sizable majority and made impressive gains in the U.S. Senate did the Obama administration begin to talk seriously about not raising taxes on Americans at the start of 2011. Until that moment, many Democrats were arguing in favor of a set of onerous burdens, including a dramatic increase in every tax bracket, from the lowest to the highest, and a dramatic increase in the marriage penalty, which would punish families who are trying to build a life together. The child tax credit was set to be cut in half, from $1,000 to $500; even teachers would have faced losing the ability to deduct their classroom expenses. And as I write this, in the late fall of 2010, millions of middle-class Americans may still face the Alternative Minimum Tax, which was originally designed just to make the superrich pay more, but has never been indexed to inflation. As a consequence, more and more working families are being forced to absorb the same tax penalties as multimillionaires. We are also seeing new burdens being placed on businesses, including a crazy requirement, which was rammed through with the health-care bill, for small businesses to report nearly every transaction on a separate government form. Only after Republican electoral gains did President Obama begin to signal that he was now open to changing this crippling requirement. But the tax debate only scratches the surface—we need to do much more. Throughout 2010, Harry Reid's and Nancy Pelosi's inability to make economic issues the top congressional priority was the most frustrating thing to me as a new senator. To maintain its own wasteful ways and bloated spending, our own government is content to further squeeze the many families and individuals, small-business owners, and double-shift workers who are just trying to get by. This has got to end.

Lowering tax rates encourages citizens to invest and to save and also to spend. It encourages businesses to hire new workers. It encourages entrepreneurs to start new ventures. Each time a business hires a new employee, we expand the productive base of our economy. We can solve a significant portion of the federal government's fiscal mess by growing our economy. A robust economy automatically creates more tax revenue—and no one has to raise taxes to do it. It's created by more jobs, more sales, more profits. It's created by more people prospering. But that fundamental economic lesson seems to have been forgotten in tax and spend Washington.

Yet even with all these taxes, we are passing on an unfathomable amount of debt to our children and grandchildren. Nearly every minute, we add a new $1 million in debt. Each U.S. citizen owes more than $43,000 in debt racked up by the government; each individual taxpayer is on the hook for over $120,000. By 2020, our national debt will be 67 percent of all the money our country makes each year. I believe this level of government spending, where the taxpayers are stuck with the bill, is immoral.

Although I am grateful that public assistance was available for brief periods for my mother, I am also deeply grateful that she never became dependent on it, that she always returned to work and also taught my sister and me the value of a strong work ethic. Today, our government has become too dependent on taxing its citizens and is always asking them for more, as if it were entitled. We need to force government to become more efficient and to ask for less.

I believe that power concentrated in the hands of one political party, as it is in Massachusetts, leads to bad government and poor decisions. That holds true for Democrats and Republicans. It is something that we need to guard against. There is little uglier than the arrogance of power or the bashing of our opponents for partisan and selfish ends.

I believe in a strong military that will protect our interests and ensure security around the world, and in a vigorous homeland defense. We owe the men and women of our Armed Forces our complete dedication and unmeasured devotion. They give us nothing less every day and every hour of our lives. If we can't support them 100 percent, we should not ask them to give their all. And we owe their families a debt as well. We can enjoy our comforts because they do without.

I believe that Americans are a dynamic, vibrant

people and nation. We are a force for good at home and abroad. I want it to be easier for the many daring and creative people among us to unleash their entrepreneurial talents. I want businesses to grow, families to flourish, and new ideas to be celebrated in laboratories, factories, offices, schools, and homes around the country. To that end, I want our government to be a help, not a hindrance; I want We the People to come first.

I have always thought that being in government service is a privilege, not a right. The Founders' ideal of a citizen legislature is that our representatives should live the same lives as the people they represent, our citizens.

Not too long after I arrived in Washington, Fred Thompson, whose house I had briefly rented, invited me for dinner at the Palm Restaurant, a far cry from my usual haunts—family-style Italian restaurants in Boston's North End. Gail and I went with Fred and his wife, Jeri, and another couple. When we arrived, I was told that the Shrivers were there and had asked whether I would like to meet them. The Shrivers are, of course, a well-known Maryland Democratic family and cousins of the Kennedys. Eunice Kennedy Shriver, who founded Special Olympics, was Jack, Bobby, and Teddy's sister. But Eunice died in 2009, a few weeks before her brother Ted. In my mind, I pictured a group of younger people, and in fact there had been some familiar-looking faces watching me from the bar when I came in. I said no, thanks, I don't want to interrupt; the six of us sat down to our meal.

When we finished, I looked over and noticed two guys standing by a door with shiny pins on their lapels. I'm a big pin guy, since in the military, pins have all kinds of significance and indicate all types of classifications. I went over to ask about their pins. They told me they were bodyguards for Governor Arnold Schwarzenegger. At that moment, I wasn't thinking of Arnold Schwarzenegger, California governor, husband of Maria Shriver. I was thinking of Arnold the movie star, *Terminator, Total Recall, Kindergarten Cop,* and *True Lies.* And that Arnold was also, like me, a former *Cosmo* guy. Can I meet him? I asked. The two guys looked at my pins and said, "You're a new United States senator; you can do pretty much whatever you want." They motioned to the door to a private room. The door was halfway open. Inside was a long table, and Arnold was seated at one end, leaning back in his chair, puffing on a cigar. I walked in, eyes focused on him, and said, "Excuse me, Governor, I'm—" And then suddenly I heard another voice, a woman's voice: "You're the new United States senator from Massachusetts, Scott

Brown." And then it hit me: these were the Shrivers, and Arnold's wife, Maria, was the one now speaking to me. They were the ones who had invited me to come over, and I hadn't realized and had declined.

I walked closer to the table, and noted that Arnold is a bit smaller in person than I had expected for a man who battled bad guys in the movies and was Mr. Universe; face-to-face, he's compact, handsome, and confident. Maria added, "Thank you for your gracious words about Uncle Teddy on election night. The family appreciated that very much." Indeed, out of respect, the first call I made on election night was to Ted Kennedy's widow, Vicki, and I kept a picture of Ted Kennedy on the mantel in the reception room of 317 Russell, my first U.S. Senate office and the office where he used to serve. But now, like a kid in a candy store, I was getting to meet Arnold, the action hero turned governor. We exchanged a few words, and then one of the other Shrivers at the table asked, "So where's your office? Are you out in a trailer or something?" I answered that I was in the same office Ted Kennedy had occupied. He looked at me with utter shock and said, "You have Uncle Teddy's office?" "Yeah, I have Uncle Teddy's office," I replied, and then Maria said, "*You* have Uncle Teddy's office?"

At this point, Arnold interrupted them all and began

to laugh. He said in his perfect Terminator voice, "Maria, Maria. It's not Uncle Teddy's office. It's the people's office." He leaned back and let the laughter roll, although no one else in the room was laughing, and I decided that now was probably a good time for me to quickly make my way to the door.

But it was a great joke. And it is also the truth.

Chapter Eighteen

SPIDER'S WEB

My life is like a spiderweb, each piece integral to the construction of the whole. A spider begins with a long, fine, sticky thread that it releases from the tips of its spinners and allows to drift on whatever breeze happens by. The edge of the drifting thread will then land and attach itself to some harder surface, like a tree branch or the quiet corner of a room. Only then, when that first line is secure, will the spider walk along its anchoring thread and add a second, like a structural beam, for extra strength and hold. Next, the spider branches out, spinning its silks wider, methodically adding radial threads and then circular threads to fortify the web at its center. The final creation is an intricate combination of perfectly positioned sticky threads for construction and also for hunting prey, and a series of nonsticky threads across which the spider moves, suspended in the air, gliding over its own design. Sever just one key link and the entire web succumbs to capricious winds. With one cut, the entire web is razed.

My life is like that web. I cannot imagine any piece of its design to be any different; I would not change any part of the experiences that have been woven together to create the larger whole.

I would not change my parents' marriage or their divorce, or the years I spent imagining but never really spending time with and getting to know my father. I would not change the men my mother married, even the bruises left by their fists and stubby fingertips. I would not change the days I went hungry or the graying, too-small clothing that after hundreds of washings I could no longer get clean. I would not change the afternoons I stole food or the day I stole those record albums. I would not change the mornings I waited for my father, when he didn't come. I would not change the fights with my mother or the times when, in desperation and devastation, I ran away. I would not change the jobs I worked at, from cleaning the deep-fry grease and the bathrooms at Dunkin' Donuts to being a liquor store clerk, a babysitter, or a house painter, or cleaning up vomit in dormitory stairs, or modeling for *Cosmo* and afterward. If I changed any one of these things, it would change the architecture of my life, and I would no longer be the person that I am today.

Had I not grown up hungering for a family, I might never have appreciated the one I am blessed with now. Had I not yearned and longed for my own dad, I might never have been the dad who was determined not to miss a basketball game or who would treasure getting a handmade Father's Day coupon entitling me to a night at the movies with my girls. Had I not known violence, not seen what it did to my mother and my sister Leeann, I might never have represented other women in their divorce cases and fought to free them from abusive homes. Had I not been forced to become a protector, of my mother and my sister, I might not have grown up to be a problem solver, to look for the way out, for the resolution to the dispute rather than focusing on the obstacle. Had I not known privation and adversity, and the fear of having barely a dollar to my name, I might not appreciate how precious each dollar is to the men and women who earn it, working at two jobs, trying to provide for themselves and those who depend on them.

Had I grown up with abundance or even in relative comfort, I might not have been nearly as hardworking and hungry myself. Had I not been preyed upon as a child, I might never have fought against predators as an adult in the Massachusetts State Senate, to try to ensure that other kids never had to face down an attacker along a wooded path or in an empty bathroom stall.

Had I never posed for *Cosmopolitan* magazine and modeled, I would have chafed for years under a different kind of dependence, owing vast sums of loans, indebted to whatever institution had written my tuition checks. Those photos gave me independence, financial and personal. Also, without them, nothing might have begun to bridge the gaping distance between my father and me. I almost certainly would never have met my beautiful Gail, and we would not have our two wonderful daughters. And because of what I missed in some parts of my life, I had the good sense to take guidance in other places, from the coaches, teachers, drill sergeants, adults, and friends who offered it. From them, I began to learn the value of devoting oneself to others, whether they be teammates, colleagues, fellow Guardsmen and soldiers, or friends and family, or complete strangers who need a helping hand.

The hardships I have known have led me to appreciate the life that I have been able to build. I can and do appreciate every moment because I know that there was no certainty when I started that I would ever reach the point where I am now.

In telling my story, I don't want anyone's pity or sympathy vote. I'm not a woe-is-me guy. Nor do I believe that personal difficulty is in any way essential for developing compassion, or dedication, or success. Instead, I believe that we each create our own playing fields and that we are all capable of overcoming whatever challenges might otherwise hold us back. The ability to persevere, whatever the circumstances, lies in each one of us.

Like a fractured bone, I have knit back stronger in the broken places. And the things that I endured are relatively small things compared with the experiences of the men and women in our military who enter combat zones, knowing that each day could be their last—or compared with the experiences of firefighters who race into burning buildings or police officers who instinctively keep one hand poised above their holster when they make a suspicious traffic stop. They put their lives at risk for total strangers; they put danger in the forefront of their duties so that the rest of us can place it in the back of ours. Fifty-one years later, I have my limbs, I have my life, and I am grateful. I am, when I look around, blessed.

The life I have led has, I hope, given me perspective, shown me the value of a second chance, of working for a greater good. Like those gossamer links of the spider-web, each facet strengthens and reinforces the others.

When I ran for the state representative slot, I prevailed on my parents to come down to the Wrentham area and hold signs for my campaign on Election Day. They came again during my state senate race, to hold signs on those final weekends and on Election Day. My dad also sent fund-raising letters to help me come up with the volume of cash needed for that special election run and helped get a lot of signatures. He had been a successful and well-respected selectman in Newburyport for sixteen years, and he had made some political contacts. He understood the demands of local retail politicking. On election night in 2004, he was there with me and so was my mom. She was less than pleased to see my father, just as she had been less than pleased to see him when I won for state rep. In her view, she had done all the heavy lifting during my growing-up years, and now he was here to reap the rewards of my success after having gone missing for so long. To her, he simply did not belong. But I told her that he is my dad, adding, "I want both of you here. I want both of you in the pictures. I want to share the night with both of you."

I can't say that I have completely put the past behind me. With my mother and my father, I retain the sense that perhaps the other shoe will drop, that there may be some other disappointment or obstacles. For my own protection, out of instinct, I keep holding a little piece of myself in reserve. Giving them my complete trust is something that I still have to work on myself. But that doesn't mean that I cannot forgive them. I have and I do. They each made some choices, whether out of necessity, emotion, foolishness, or inexperience, which I know they now regret. My mother had every right to be angry and hurt, given all that she had gone through. My father had his own family turmoil to overcome. At a certain point, as adults, we stopped dwelling so much on our pasts and focused on our future. I'm sure this book will be hard for them to read, that it will be hard, painful even, for them to relive those early years and those parts of our lives. But I hope they are proud too of the distances that we have traveled and the steps toward healing that we have all taken, many of them now together.

My father tells me that he loves me. He calls me and my family; he wants to be in my life, to see me, to spend time with me, with Gail, and with Ayla and Arianna. He is there for me; and for that I'm thankful. I am the one now who parcels out each visit—my bruising schedule and the biweekly flights to and from Washington, D.C., barely leave me with time for an occasional evening with my wife, or a quick meal with my daughters. The wheel has turned in ways that none of us expected; now that my parents have time for me, my time is no longer my own. But my mother is getting a second chance with her four grandchildren, happily attending Ayla's basketball games and singing events, applauding at Arianna's horse shows or horse races, and cheering at the games of Leeann's daughter and son. Now, without the pressure of being a provider, she can take pleasure in the lives of her children and her grandchildren.

Indeed, if my life is a story of second chances, so are my parents' lives. Not long after we began to reconcile, my father started volunteering for the Jimmy Fund, one of the oldest health charity organizations in the nation. The Jimmy Fund works with the Dana-Farber Cancer Institute in the battle against cancer—in the summer of 2010, I joined five thousand cyclists, including Senator John Kerry, to ride in a charity race on behalf of the fund. I dedicated my ride to one man, Judge Samuel Zoll, who has been fighting cancer himself.

My father has helped raise over $100,000 for the Jimmy Fund, to help friends and neighbors and people he has never met. He has walked in parades, carrying a blanket, so that bystanders could toss stray coins for donations. In 2010, I was able to present him with a special award to mark his twenty-five years of service. Today, it is with great pride that I stand anywhere, whether on a podium or in a backyard, and introduce my dad or my mom.

In my grandparents' final years, my mom moved back to New Hampshire to be their caretaker. My grandfather lived long enough to see me married and to see Ayla and Arianna born. He died in 1994, before I became a Wrentham selectman. My grandmother lived until 2001, and Gail, the girls, and I tried as often as we could to make the drive to New Hampshire to see her. Gram is buried beside my grandfather, in Portsmouth, in a green, grassy cemetery, a little ways away from the water and the bustle of downtown. I miss them still.

My mother made their remaining time as comfortable as possible, and she did it with love. Afterward, she stayed in New Hampshire, caring for a couple of houses that my grandfather had passed down, making a new life for herself. She has become a dedicated volunteer at local animal shelters, caring for pets and others in need. At last, she seems to have found her home.

And she can know that there were many things she did right. The powerful work ethic that my sister and I share was in many ways taught to us by our mom, who never shirked from doing whatever job she could find to support us, who never expected a handout or help. No matter what my mom's experiences were, both Leeann and I have rich, loving marriages. And we have learned from our own growing up. We have not repeated the same mistakes that were made in our home; instead, we've been made wiser by them. We've consciously tried to apply the lessons of our lives, good and bad, and do things differently. We are determined to show our own children that they can get past whatever difficulties they face in life, in marriage, and in their family. We have tried to teach them how to choose love wisely and how to stay in it for the long haul. As I look at my daughters, at Leeann's son and daughter, and at Brucie's and Robyn's kids, I believe that this next generation is more resilient than our own, that the fruit of each new generation is better than the one before it.

When I decided to run for the U.S. Senate, I told both of my parents before I announced, and, from the beginning, they both wanted to help. This time, there were no barbs, no recriminations, no simmering hostilities, just everyone working together toward one goal. My mom asked all her old high school friends to volunteer. She asked for donations, and she did whatever she could. My dad went out and got hundreds of signatures for my petitions to get on the ballot; he scouted out campaign sign locations; he made phone calls and worked to help with fund-raising. And quickly, it spread beyond my parents. Leeann and her husband, Rich, came down from New Hampshire to find the best sign locations, Robyn and Bruce chipped in however they could. My dad's sister, my aunt Linda, did her part. Her daughter, my cousin Shannon, took most of our campaign photographs; she captured amazing images at hundreds of campaign stops and also went door to door for me. Gail's mom, Anne, got hundreds of signatures, and her sister, Jenny, was great at working the phone banks and finding sign locations. They all pulled together without complaint. They called to check in on me, to offer whatever help they could. At one point during the race, I said to Gail, "We have one of the most dysfunctional families in the world, and who would have ever thought that my running for the U.S. Senate would be the thing to pull everyone together?" I was amazed at how they all gave their time and their dedication to try to help me win. It was gratifying and humbling, and it gave me hope, not only for the election, but for our family.

On election night, my entire extended family was standing there with me on the platform. It was so packed that people could have fallen off. It was our first true family reunion, as it were, with all these intersecting threads and storylines coming together in one place. And everyone was just happy to be there, on that one stage. There was no bitterness, just shared pride and joy, mine perhaps the greatest of all, as I glanced around. My father was behind me; my mother was next to Arianna; Gail and Ayla were on either side. I treasure that moment every bit as much as the win they had helped me to achieve.

Not long after I won the election, Massachusetts Congressman Barney Frank, with whom I've since developed a cordial working relationship in our role as colleagues from the same state, said a bit dismissively of me, "Having an old truck and two daughters are not usually policy arguments." I actually take it as a compliment. I am happy to be measured by my family; if my life were defined by the worth of Gail, Ayla, and Arianna, it would be immeasurable. These three wonderful women are my rock of stability. We are always there for each other. I take the greatest pride in Gail's talent as a journalist, in her stellar reputation for always being fair, accurate, and extremely hardworking. She is incredibly accomplished and skilled. What most people don't often get to see is the deep commitment Gail and both our daughters have to others.

Gail has volunteered at, donated time to, and emceed

more than a hundred charity events over the course of her career. A couple of years ago, Gail and Arianna traveled together to Ciudad Juárez in Mexico with our local church to help build an orphanage. Arianna, who loves to draw and paint, was asked to paint images of local reptiles and lizards to decorate the children's rooms. Arianna is marvelous with animals. For years, she has gone out into the yard, out into the woods, and animals simply follow her home. We now have two dogs and a cat. Over the last decade, we've had an iguana, a Quaker parrot, a box turtle, hamsters, a freshwater and saltwater fish tank, Chinese fighting fish, and at different times, four separate horses. Arianna has developed an amazing love for and intuition for horses. She can walk up to a horse, look it in the eye, and calm it down. Her ability to read animals and also people, to sense their needs, is a gift. And while she has occasionally modeled, like her mom, she is planning on becoming either a veterinarian or a doctor, and earned close to a 4.0 GPA during her freshman year in college. Ayla too is unfailingly generous. She has sung at countless charitable events. A few months after she reported on a young girl with cancer for the CBS *Early Show*, she called the girl and her mom up and invited them to come to visit the beach with her for the afternoon. She didn't do it for a camera or a spotlight; the story was long since over. She did it because she cared about them and wanted, on her own, to do a little bit more, to make a difference.

Please measure me by my family. I can't take the credit, but I couldn't be more proud.

On Father's Day 2010, a few weeks after Ayla's college graduation and the conclusion of Arianna's freshman year, and my first Father's Day as a U.S. senator, I decided that I really wanted to ride my bike.

I have a theory that part of Father's Day should be a chance for a dad to do whatever he wants. On that day, I was in Rye, New Hampshire, at our little weekend place, with my wife and daughters. I had talked to my mom and was going to see my dad. Arianna, Ayla, and Gail had planned a special surprise for him, a new golden retriever puppy to replace the one that had died. But I also really wanted to ride my bike.

So I decided to ride down along the New Hampshire coast, through the twists and curves, past the short sand beaches and faded black rocks, to the Massachusetts line and down into Newburyport and out to Plum Island. That stretch of road is one of my favorite rides in a car, but after all these years, I had never done it on a bike. And as I sat on my bike, each leg rising and falling in one fluid motion, my lungs working in time with my legs, my back curved low over the handlebars and the wind passing above me, the sound of its rush overwhelming my ears, I was content. The water rolled past, waves drifting in and out, and each noise had its own perfect rhythm. I could be alone with my thoughts, I could dissect problems in time to the motion of my body and the sway of the tide. When a drenching afternoon rain came, washing over me and the road, I was at peace.

In these moments, it is hard not to feel close to God, to be able to ask for guidance, forgiveness, and strength, to be able to confess any doubts or fears, and to be able to give thanks. And on this day, I was thankful. As I rode, I thought back to those other rides, thirty-five or thirty-six years ago, rides of desperation and rides of fear, when I attempted to escape from Salem Street in Wakefield and wind my way through the hills to Newburyport. Those rides were made in anger, frustration, and even grief for the life that I wished I had. They were rides to run away, to leave everything I knew behind.

Now, I find myself riding with a far different purpose. I'm riding to go see my dad. I'm riding to see the smile on his face when the new puppy bounds out of the car. Behind me in the car are Gail, Ayla, Arianna, and the puppy. They left about twenty minutes after me, and I beat them to my father's house by ten minutes. I checked it on the clock. Even though it was Father's Day, and I'm now past fifty, I'm still competitive.

And I also know that you go much faster and the journey is much sweeter when you are not riding away from something, but when you have a hopeful destination that you are riding toward.

ACKNOWLEDGMENTS

My deepest thanks go to my wife, Gail, and my daughters, Ayla and Arianna. They have been a special blessing in my life, a family that I treasure. I am grateful for their faith in me, for their support, and for their constant love. Gail, thank you for all the days and hours that you have been by my side. It has been more than a quarter century since we met, and when I hear your voice, I know that I am still one of the luckiest men on earth.

My thanks also to my mother and father, and to my sister Leeann, and my other siblings, Robyn and Bruce. I know that parts of this book will be difficult for my parents to read, but they have also proved that the past does not have to determine the future. As adults, we have rediscovered each other, and I am grateful to have them in my life. Leeann, thank you for sharing your memories and your support and encouragement, which have been invaluable. And thank you, Robyn and Bruce, for our bonds.

As soon as I won the election in January 2010, publishers began calling, interested in a book. I had never considered such a project, but the more I thought about it, the more I wanted to tell my story, the good and the bad. My hope in sharing my life is that it will give hope to others, that other people who are struggling will be reminded that things can get better.

I want to specially thank Lyric Winik, a terrific and very gifted writer, for the many hours we spent together and particularly for her patience. She listened to me, guided me, and gave me all the time I needed to open up, to talk about the many parts of my life, helping me to thoughtfully convey my experiences, challenges, and successes, as well as to portray the people at the core of this story. Thanks for the care she gave to help put my story into words. Thanks also to her husband, Jay, and their sons, Nathaniel and B.C., who graciously shared her for countless long workdays and never complained about my frequent dinnertime phone calls.

Bob Barnett of Williams & Connolly was vital in

making this book project come to pass. I am most appreciative of his dedication and wise and excellent counsel.

At HarperCollins, I wish to thank publisher Jonathan Burnham for his strong support and commitment to this book. Executive Editor Tim Duggan has been an outstanding advocate, and it has been a true pleasure to work with Jonathan and Tim. My thanks to them for having such faith in my life's story. I would also like to thank other members of the HarperCollins team for their great work on behalf of this book, particularly Tina Andreadis, Chris Goff, Tom McNellis, and Allison Lorentzen.

My heartfelt thanks to the many people who shared their recollections with me, most especially Judge Samuel Zoll, and my coaches—John White and his wife, Cathy; Ellis "Sonny" Lane and his wife, Paula; and Brad and Judy Simpson—as well as Bob Moore. My debts to them over a lifetime are too great to repay. I would also like to warmly thank some of my friends who were always willing to spend some time reliving our pasts, no matter what day or hour I called: Mark Simeola, Jimmy Healy, Mike Quinn, Bruce Cerullo, Bobby Rose, Pabs, Lana, Alb, Turner, Bob Najarian, Billy Cole, Gonzo, and Dave Cornoyer. To paraphrase the poet William Butler Yeats, I can truly say, "My glory is I have such friends."

My special thanks to Eric Fehrnstrom, Peter Flaherty, Beth Myers, and Gail Gitcho for taking their private time to read the manuscript and to offer their very thoughtful and extremely helpful comments. Thanks to Dan Winslow and Liz Figueria for their excellent legal work and key contributions.

Under impossible deadlines, Jo Shuffler was invaluable in producing transcripts that were vital to the manuscript, and she always did it with good cheer. Thanks also to Shannon Power, my cousin and a wonderful professional photographer, for following me on the campaign trail and so generously sharing her photos. And also to Jenny, Paul, Cindy, Anne, Jonathan, and Callie, my Zeta Psi brothers, and Tufts and BC Law buddies, as well as the Monadnock Sportsmen's Club, the Bay State triathlon team, and my many friends in Wrentham. Thanks to North Attleboro Selectman John Ryhno and his wife, Sherri, for all their help over the years on the campaign trail. And to State Senator and Mrs. Richard Ross. My thanks as well to Major General Joseph Carter of the Massachusetts National Guard for his leadership. Also to Mitt Romney and John McCain for their support when it most mattered. And to Beth Lindstrom and so many others who were part of the special election campaign team.

I'm very appreciative of the hard work of Steve

Schrage, a brilliant policy analyst and a wonderful guy, who stepped in to run my U.S. Senate office from its first days through the November 2010 elections. My thanks also to Greg Casey for his assistance in Boston and D.C. and to my entire D.C. and Boston U.S. Senate team for dealing with the vital and challenging issues that face our nation and the people of Massachusetts every day.

In the U.S. Senate, I have benefited enormously from the thoughts and friendship of many of my colleagues. I would particularly like to acknowledge Mitch McConnell, Jon Kyl, John Thune, Mark Warner, Orrin Hatch, Richard Burr, Lindsey Graham, Bob Casey, Tom Carper, Mark Udall, Olympia Snowe, and Susan Collins.

I would like to add a personal note of thanks to Sister Katie and the nuns of Mount Saint Mary's Abbey in Wrentham. I came to know them when they needed some assistance while I was in the state legislature, but I found them in so many ways to be an example to me. My family and I have been uplifted by their prayers; my daughter Ayla has gone to work and sing with them in the fields when they harvest their summer crops. Although I am not a Catholic, I am grateful to have been touched and welcomed by their deep faith.

There are also several books that were particularly

useful in preparing this book, and I'd like to acknowledge them and their authors: *History of Middlesex County, Massachusetts,* by Samuel Adams Drake, Estes and Lauriat Publishers, Boston, 1880; *Massachusetts, A Concise History,* by Richard D. Brown and Jack Tager, University of Massachusetts Press, Amherst, 2000; *The Boston Irish, a Political History,* by Thomas H. O'Connor, Back Bay Books, Boston, 1995; and *Common Ground,* by J. Anthony Lukas, Vintage Books, New York, 1986. In addition, in reviewing events from the 2009 to 2010 special election for the U.S. Senate in Massachusetts, articles that appeared in the *Boston Globe,* the *Boston Herald,* and the *Sun Chronicle* were particularly useful, as were archived radio and television footage. Thanks to Dan Rea's show *Nightside* for supplying an audio file of the 2009 election forum.

Finally, I would especially like to thank the people of Massachusetts and around the nation who have offered their support, not only during the election, but in the year that has followed. I am humbled and honored by the chance to serve you and our country. I also hope that some of the readers of this book will decide that they too can go out and make a difference. Please get involved, join a team, help coach some kids, run for local office. In my life, the people who have done just that have made all the difference.

Except as noted below, all photos are courtesy of the author. Grateful acknowledgment for permission to reproduce photos is made to the following:

Courtesy of Fay Foto Service: insert page 4, top and bottom.

Courtesy of Tufts University: insert page 5, top and bottom.

Courtesy of Ed Bennett: insert page 6, bottom.

Courtesy of Cathy Copeland, Cathy Copeland Art and Design: insert page 7, top.

Courtesy of Jean Renard, Phil Porcella, Robert Kaufman: insert page 7, bottom.

Courtesy of Mark Kasianowicz: insert page 11, bottom.

Courtesy of Capstone Photography: insert page 12, middle and bottom.

Courtesy of Shannon Power (www.shannonpower.zenfolio.com): insert page 13, all photos; page 14, bottom; page 15, top and bottom.

Courtesy of AP/Steve Senne: insert page 14, top.

Courtesy of AP/Winslow Townson: insert page 14, middle.

Courtesy of Mark Stockwell: insert page 15, middle.

Courtesy of UPI/Matthew Healey: insert page 16, top and middle.

Courtesy of Brendan Hoffman (www.brendanhoffman.com): insert page 16, bottom.